Your PACKET Companion

By Steve Ford, WB8IMY

Published by: **The American Radio Relay League**
225 Main Street, Newington, CT 06111

Foreword

Packet radio has sparked an Amateur Radio revolution. Yet, despite its remarkable power, it is one of the easiest modes to operate. All you need is a transceiver, a computer—or a simple data terminal—and a device called a TNC. That's it!

By communicating with automated packet bulletin boards, you'll be able to keep your finger on the pulse of the Amateur Radio community. You'll also be able to send messages to other hams around the world. You can even use your packet station to enjoy real time person-to-person conversations.

Packet is constantly evolving. Innovative networking systems are as close as your keyboard. Packet satellites can forward your messages around the globe, or send images from space that you can see in your own home.

In *Your Packet Companion*, Steve Ford, WB8IMY, introduces you to the fascinating world of packet radio. Every aspect of packet, from bulletin boards to satellites, is discussed in easy to understand terminology. You'll learn how packet works, how to assemble your own packet station and how to use packet to your best advantage.

If you haven't tried packet radio yet, I urge you to do so. You'll be doing more than having fun; you'll be taking your first steps into the future of Amateur Radio.

David Sumner, K1ZZ
Executive Vice President
May 1992

Acknowledgments

This book is more than the effort of one individual. Others made valuable contributions by supplying information and encouragement. An incomplete list includes: Jeff Bauer, WA1MBK, Jon Bloom, KE3Z, Joel Kleinman, N1BKE, Luck Hurder, KY1T, Bill Tynan, W3XO, Brian Battles, WS1O, and Jim Kearman, KR1S.

Steve Ford, WB8IMY
Assistant Technical Editor

Contents

CHAPTER 1

What is Packet Radio?

et's say that I wanted to transmit the contents of this book from my computer to your computer. I could establish a radio link to your station and use it to send everything to your computer in one transmission. It sounds easy so far, doesn't it?

Why don't we make our example a little more challenging? We'll add two more hams on the same frequency. These guys don't want to send books to each other, but they *do* want to swap lengthy opinions on the latest music. I'm quick on the trigger, however, so my station transmits first. The entire text of this book is suddenly flying across the airwaves in a long, continuous stream of data. My signal is occupying the frequency, so the other hams have to stand by until I'm finished.

At long last my transmission ends. Now the other hams can send their information. But wait! During my transmission there were bursts of noise and interference that destroyed some of the data. You can't have a book with errors, so there is only one thing I can do: send the *entire book again*! Without hesitation, my computer switches on the transmitter and

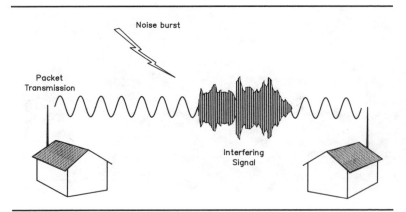

Fig 1-1—Every data transmission is subject to losses caused by noise and interfering signals. The longer the transmission, the more likely it is that errors will occur. Lengthy transmissions also tie up frequencies, preventing other stations from sending information.

repeats the outrageously long transmission.

By this time everyone is staring at their computer screens and muttering, "This is ridiculous. There must be a better way."

Of course, they're right!

Making a Big Job Smaller

Let's use the same example, but this time we'll use *packet* techniques to get the job done. The first step is to establish a *connection* between our stations. We say our stations are "connected" when the packet transmission and reception *protocols* are active and we're ready to transfer data. (A protocol is a formal, standardized way of doing something. Clubs often use Robert's Rules of Order to keep meetings civilized. Robert's Rules of Order are a set of protocols.)

My station begins by digesting the text of the book and constructing a number of "byte-sized" packets for transmission. The packets are lined up in proper order like airplanes sitting on a taxiway waiting to take off. Within a fraction of a second, the first packet is on its way.

Your station receives the packet and checks it for errors. In the meantime, a timer has started back at my station. If I don't hear from you before the timer reaches zero, the packet is sent again. For the moment we'll assume that my first packet arrived error-free. Your station communicates this fact by sending a special signal known as an *ACK*, or *acknowledgment*. When I receive your ACK, my next packet is transmitted.

"Hold on!" you say. "What happens if your station doesn't hear my ACK for some reason?" If your ACK arrives distorted and unreadable—or not at all—my timer simply

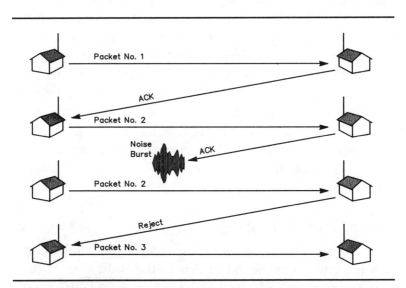

Fig 1-2—AX.25 packet protocols provide an efficient means to transfer information via radio. In this example, *packet #1* is sent and acknowledged (ACK). *Packet #2* is sent next, but a burst of noise blocks reception of the ACK signal. After a certain amount of time has passed without an acknowledgment, *packet #2* is retransmitted. Since the receiving station already has this packet, it sends a REJECT and *packet #3* is transmitted.

continues to count down to zero. Eventually, my station will resend the first packet. When you receive the first packet again, your station will, in effect, say, "What's going on here? I already have this packet. Send the next one!" It indicates its displeasure by transmitting what is known as a *reject frame*. When I receive the reject, my station will automatically send the next packet.

Through this process of timers, ACKs and rejects, all of my packets will arrive at your station in one piece—and vice versa. By sending the data in small packets, it's possible to communicate under conditions ranging from ideal to marginal. When signals are strong and clear at both stations, the packets are transmitted and acknowledged in rapid-fire sequence. As the signals degrade and noise or interference worsens, the data transfer rate becomes slower. (Acks are lost, packets must be retransmitted several times and so on.)

During the silent periods between our transmissions, the other hams can send their packets, too. Packet is extremely patient. If a station has a packet to send and another station is transmitting on the frequency, it will wait until the transmission stops before trying to send its data. Packet is also very persistent. A packet will be transmitted and retransmitted many times before the system finally gives up and breaks the connection.

A Bit of Packet History

Packet is relatively new to Amateur Radio, but its roots reach back to experimental efforts in the '60s. One of the earliest ancestors of packet radio was *ALOHANET*. *ALOHANET* was created in 1970 as a wireless network of university computer systems, each on separate Hawaiian islands.

The lessons learned from *ALOHANET* were applied to wired systems as well. Large computer networks needed an efficient method to communicate with many terminals—some in distant locations. Telephone lines were the preferred medium, but even they had a tendency to be unreliable at times. In addition, there had to be a way to allow many users to "talk" to the same computers over the same telephone lines without frustrating delays. (Think about our previous example and imagine hundreds and even thousands of hams trying to send data on a single frequency!) As you can probably guess, packet provided the solution. A commercial packet protocol known as *X.25* was soon developed.

The Personal Computer Arrives

Before 1975, the idea of owning your own computer seemed ludicrous. Computers were highly specialized, expensive devices that dwelled primarily in office buildings. Even if you had the space in your home to accommodate the machine, you had to be independently wealthy to afford it! A few computer hobbyists (known as *hackers*) were assembling small models that could perform limited tasks, but the barrier had yet to be breached between the world of the computer and the world of Amateur Radio.

The first cracks appeared in 1975 when the Altair computer became available in kit form. It was a pain to program and its capabilities were few. Even so, it placed the power of a true computer in the hands of radio amateurs for the first time.

Soon thereafter, the first Apple appeared along with the Commodore PET, the Radio Shack TRS-80 Model 1 and a host of other personal computers. Hams snapped up these new machines right and left. Programs were written for various applications including logging, QSL printing and even CW and RTTY. Visionary programmers knew that this was just the tip of the iceberg. Personal computers could do more than this—much more!

The Birth of the TNC and Amateur Packet Radio

In Canada, hams began to experiment with the idea of using commercial packet protocols for error-free computer-to-computer communications over radios. In 1978 the Canadian Department of Communications legalized the use of packet radio for Canadian amateurs. Even before the announcement was made, Canadian hams were moving forward with designs for a special interfacing device called a *terminal node controller*—or *TNC*. The TNC accepted data from a computer or data terminal and assembled it into packets. In addition, it translated the digital packet data into special audio tones that could be fed to a transceiver. The TNC also functioned as a receiving device, translating the audio tones into digital data a computer or terminal could understand. (The Canadian TNCs did not include modems as part of their design. The American-made TAPR TNC-1 was the first TNC to incorporate a modem.)

If you're saying to yourself, "These TNCs sound a lot like telephone modems," you're pretty close to the truth! If you're familiar with so-called *smart* modems, you'd find that TNCs are very similar. TNCs contain microprocessors, memory and their own software. Very smart devices indeed!

Packet radio experimenters in the United States got their break when the FCC approved packet for Amateur Radio in 1980. By 1981, the experimenters had agreed on an amateur packet protocol modeled after X.25. They called it *AX.25* (for *Amateur* X.25). By 1985, it had become the worldwide standard for Amateur Radio packet communications on all bands, including HF.

The timing for the introduction of packet radio was perfect. Cheap personal computers were flooding the market and Amateur Radio software was becoming more sophisticated. Many of the first packet TNCs were sold as kits by organizations such as TAPR (the Tucson Amateur Packet

This is the TAPR TNC-2, one of the most popular TNCs in packet-radio history. Many of the commercial TNCs available today are direct descendants of the TNC-2.

Radio Corporation). Commercially manufactured TNCs would soon follow. They were expensive at first, but market demand and competition quickly drove prices to affordable levels. Within a few years, multimode communications processors (MCPs) appeared, offering packet, RTTY, AMTOR, CW and several other modes in a single device.

As more amateurs became packet-active, local packet networks sprang up throughout the country—primarily on the 2-meter band with FM as the preferred mode. Packet bulletin board systems (PBBSs) formed the nucleus of most local networks. They acted as clearinghouses for information that flowed through the packet system. Bulletin boards were linked to each other through packet repeaters known as *digipeaters*. If the distance between one network and another was too great for a reliable path on VHF, packet *gateways* provided access to the HF bands for mail and bulletin forwarding. The FCC granted Special Temporary Authorities

(STAs) to a number of gateway stations, permitting *unattended* packet mail forwarding on the HF bands on a temporary, experimental basis.

With the rapid expansion of packet through the '80s, isolated local networks were blended into national packet systems. National systems soon extended their communications capabilities internationally, making amateur packet radio a global phenomenon!

What Can I Do with Packet?

For starters, you can use packet bulletin board systems to send and receive messages from amateurs throughout the nation and the world. Do you have an amateur friend in another city or state? Why not keep in touch via packet? Are you having a problem with a piece of equipment? Post a general bulletin on the PBBS and maybe someone will have the solution.

When you read bulletin board messages, you'll discover that they cover a very broad range of topics. There are "for sale" and "wanted" messages—just like the classified section of a newspaper. You'll also see messages promoting radio-controlled airplane expositions, model rocket launches, genealogy groups and so on.

In their public service role, PBBSs are often used for sending and receiving National Traffic System messages. You can use the packet bulletin board to send NTS "traffic" to non-hams, for example. (How about a "happy birthday" greeting to a friend or relative?) You can also request a list of NTS messages awaiting delivery in your area. Any ham can deliver NTS traffic and it's a very rewarding experience!

More advanced PBBSs offer magazine bibliographies (find that old article you've been searching for), call sign directories (not everyone owns a Callbook!) and many other useful features.

Jeff Weinstein, K1JW, was caught by the news cameras of a commercial TV station as he used packet radio from the offices of the National Weather Service during a Simulated Emergency Test (SET).

If you're a DX hunter, you'll love packet! Special networks called *DX PacketClusters* are popular throughout the country. By connecting to a *PacketCluster* in your area, you can find out which DX stations are on the air and where they are. *PacketClusters* can also provide the latest information on band conditions and solar activity. Are you trying to find the address for a DX station you just worked? Some *PacketClusters* can access QSL databases which list addresses and/or QSL managers for many DX stations.

Are you interested in satellites? You'll find plenty of packet activity in the heavens! At the time of this writing, there were nine amateur packet satellites in orbit. In addition, US Space Shuttle missions frequently include Amateur Radio packet stations or automatic packet "robots."

Have you considered a "private" packet link between you and several of your local Amateur Radio friends? There's ample room on the VHF, UHF and microwave bands to operate dedicated packet radio links. Imagine being able to access a friend's computer any time of the day or night! You could leave messages or even transfer public-domain software.

If you like the great outdoors, consider taking packet radio along on your next hiking or camping trip. With the proliferation of tiny portable computers, outdoor packet is a snap! It's especially useful during Field Day activities.

Can I Just Chat With Someone On Packet?

You bet! Real-time packet conversations are common on all bands. In fact, I've worked a fair amount of DX through live HF packet contacts.

During packet conversations you have the luxury of being able to compose your thoughts before you start typing. When you're talking into a microphone, on the other hand, a lapse in thought can result in some awkward ramblings. ("Well . . . ah . . . let's see. . . ")

Packet conversations are quiet, too. You can carry on a spirited debate with a ham across town and all that can be heard is the clicking of your keyboard. This is especially handy when other family members are asleep!

Do I Have to be a Computer Expert to get on Packet?

Not at all. If you can read manuals and follow instructions, you can easily become an active packeteer. In some instances, you don't even need a full-fledged computer

---

Who says you can't take your packet with you? Walt May, KA7STK, of St George, Utah, installed this portable packet unit inside a watertight Pelican camera case. It's an ideal go-anywhere packet station! (*photos courtesy of KA7STK*)

to get started. (We'll talk more about that later.)

Like any other facet of Amateur Radio, you're going to have to learn some new concepts, but that's part of the enjoyment! I guarantee that you'll make some mistakes along the way, but I also guarantee that you'll profit from every one.

Throughout this book we'll discuss the basics of packet operating and pass along a few tricks of the trade. In the process, we'll destroy the myth that packet is only for computer geniuses. Contrary to anything else you may have heard, packet is easy, affordable and, most of all, *fun*. If you exploit this mode to even a fraction of its full potential, you'll wonder how you ever got along without it!

CHAPTER 2

Assembling Your Packet Station

 o you're convinced that you want to join the ranks of the packeteers. I bet you're wondering how much this little adventure is going to cost you. At the risk of sounding evasive, the answer is: *as much as you're willing to spend.*

No kidding! There are several roads you can take to assembling your packet station—and some roads are more expensive than others. Let's take a look at the basic components of a VHF-packet station: Computer, Terminal Node Controller and Radio.

The Computer—The Brain of Your Station

As a dedicated computer fanatic for over 10 years, I always make the mistake of assuming that every ham has a computer in the shack. For the moment I'll pretend that you don't. What kind of computer do you need to operate packet?

Within the past several years, IBM PCs and compatibles have become the de facto standard in Amateur Radio. Despite PC domination, *any* computer—new or used—will function on packet if it provides a standard EIA-232-E (RS-232-C) *serial* port, or a *TTL* (transistor-transistor logic) port. Your TNC will require one or the other to talk to your machine. The

Bill Wawrzeniak, W1KKF, enjoys a packet QSO at the Meriden (Connecticut) Amateur Radio Club station, W1NRG. Their system uses a data terminal to communicate with the TNC. By using a terminal rather than a computer, the club was able to provide packet capability for its members at minimal expense.

computer you choose must also have *terminal software* available (see Table 2-1).

Terminal Software

Your TNC is a pretty smart device, but it lacks the means to interact with you, the operator. You need a keyboard to enter your commands and data. By the same token, your TNC needs a screen on which to display its responses. If you put together a monitor screen, a keyboard, a microprocessor and the appropriate data communications *firmware* (software that's permanently stored in memory), you'd have the classic *data terminal*. By using this terminal, you and your TNC could communicate with each other.

So how do you make your personal computer function

```
 🍎  File  Edit  Control  Windows  Settings  FAX
                 Send data to Stream A (N1DCS-2)
 r 31952|

                              Monitor
 MTM>N1IBR [05/08/91  22:17:58] <RR R F R2>
 KB1TH-15>KA1TDL [05/08/91  22:18:17] <I C S1 R5>:i had the split on ...
 MTM>KB1TH [05/08/91  22:18:18] <RR R R3>
 WF2M-15>WB2QJA-4 [05/08/91  22:18:18] <REJ R3>
 MTM>KB1TH [05/08/91  22:18:21] <I C S6 R3>:WHATS GOING ON OVER THERE OFF FREQ.
 KB1TH-15>KA1TDL [05/08/91  22:18:22] <RR R R6>
 MTM>N1IBR [05/08/91  22:18:32] <RR R F R3>

 ▆▆▆▆▆▆▆▆▆▆▆▆▆▆  Connection A (N1DCS-2) ▆▆▆▆▆▆▆▆▆▆▆▆▆
 *** CONNECTED to N1DCS-2 [05/08/91  22:16:12]
 haven2:N1DCS-2> Connected to N1DCS-4
 [MBL-5.14-H$]
 Welcome to Shoreline BBS - CSTN
 CT/NY/LI Mailbox and MsgSwitch
 West Haven, CT 06516
 READY >
  Msg# TS   Size TO      @ BBS   From    Date    Subject
 31956 B$   1997 IPARC  @ALLBBS DF9ED   08-May  IPARC
 31955 B$   2769 NEED   @ALLBBS DG8FZ   08-May  Icom-xx71 Ram-Comments ?
 31953 B$    844 SALE   @NEBBS  KR1ZX   08-May  2 METER CHEAPIE 4 SALE
 31952 B$   2683 AMSAT  @AMSAT  W3IWI   08-May  UoSAT-F Launch/Status, Part 1
 31949 B$    306 ALL    @NEBBS  N1GBP   08-May  Computer For Sale...
 31938 B$   2352 ALL    @USBBS  WA1PHV  08-May  May NEPRA meeting
 31937 B$    438 ALL    @USBBS  N1API   08-May  New DX Country?
 31924 B$   2399 ALL    @ALLBBS I3BIP   08-May  QSL from AP2SAR
 31923 B$   2984 ALL    @ALLBBS DG9LJ   08-May  cq dx from germany
 31922 B$    633 WANTED@NEBBS  KA1EGY  08-May  2-m All-mode
 READY >
```

Fig 2-1—*MacRATT* is an example of an Amateur Radio packet software package written for the Apple Macintosh. Other packet software packages are available for a variety of computers.

as a data terminal for your TNC? If you simply plugged your computer into your TNC, the computer would sit quietly and do absolutely nothing. After all, you haven't supplied the instructions (software) to tell it what to do. Your computer needs a *terminal program* to make it behave like a data terminal. (see Fig 2-1).

Some terminal programs are very basic. They get you talking to the TNC and little else. Others have all sorts of goodies such as packet bulletin board (PBBS) "zap" functions. Press one key and the software will automatically instruct your TNC to connect to your local PBBS. Any mail will be instantly downloaded and saved to a disk file. A list of new bulletins will be requested and the bulletins will be scanned for key words of your choice (such as "ATV," "ARRL," "AMSAT" and so on). If a bulletin contains a key word, it is downloaded and saved as well!

Table 2-1

A Packet Terminal Software Sampler

Please note that the products and addresses shown below are subject to change. Contact the distributors to verify availability before ordering.

Apple II

APR: Send a blank, formatted diskette and a postage-paid, self-addressed disk mailer to Larry East, W1HUE, 1355 Rimline Dr, Idaho Falls, ID 83401

Apple Macintosh

MacRATT: Advanced Electronic Applications (AEA), PO Box C-2160, Lynnwood, WA 98036-0918

Virtuoso: James E. Van Peursem, KEØPH, RR #2, Box 23, Orange City, IA 51041

Atari

PK2: Electrosoft, 3413 N Duffield Ave, Loveland, CO 80538

Commodore

TNC64: Texas Packet Radio Society, PO Box 50238, Denton, TX 76206-0238

DIGICOM>64: A & A Engineering, 2521 West LaPalma, Suite K, Anaheim, CA 92801, tel 714-952-2114

MFJ Enterprises, Box 494, Mississippi State, MS 39762 tel 601-323-5869

(*DIGICOM>64* is a TNC *emulation* program that causes the computer to behave as a TNC. The only external device required is an inexpensive modem interface. The software and modem can be purchased separately.)

IBM PC

LAN-LINK: Joe Kasser, G3ZCZ, PO Box 3419, Silver Spring, MD 20918. Evaluation copy available for $5.

PK GOLD Enhanced: InterFlex Systems Design Corp, PO Box 6418, Laguna Niguel, CA 92607-6418.

PC-Pakratt II: Advanced Electronic Applications (AEA), PO Box C-2160, Lynnwood, WA 98036-0918.

BayCom: PacComm Inc, 4413 North Hesperides St, Tampa, FL 33614-7618, tel 813-874-2980

(*BayCom* is a TNC emulation program similar to *DIGICOM>64*.)

Tandy Color Computer

COCOPACT: For more information, send a self-addressed, stamped envelope to Monty Haley, WJ5W, Route 1, Box 150-A, Evening Shade, AR 72532

It isn't necessary to use a terminal program written specifically for packet applications. There are many terminal programs available for digital communication over telephone lines (such as the software you'd use to access CompuServe or Prodigy). These programs will work for packet, but they may be a little awkward at times.

Terminals—Using the Real Thing!

So far we've been talking about using software to make your computer emulate a terminal. Why not use a real data terminal to communicate with your TNC?

Many amateurs take this approach to packet operating because it's very inexpensive. If you shop at hamfests or computer fleamarkets, you'll often find used data terminals being sold for very reasonable prices. (At one hamfest, I saw a fellow selling a truckload of used terminals for $10 each!)

As you've probably guessed, there are some potential pitfalls. Consider the following factors carefully before you reach for your wallet:

❑ Check the condition of the terminal. Look at the screen closely. Do you see faint, shadowy lines where the phosphor coating has deteriorated after years of constant use? (On some overused terminal screens you'll actually be able to see horizontal lines where the text was displayed!) This is a sure indicator of a terminal that's seen better days.

❑ How will you save data? Most terminals do not contain disk drives. If you download a long message from your packet bulletin board, you'd better be a fast reader. Once the text moves off the top of the terminal screen, it's gone!

❑ If you want to try other packet protocols such as TCP/IP (which we'll discuss later), how will you run the software? Remember: a data terminal is *not* a computer.

Used data terminals are excellent if you want to get started in packet while keeping your costs to a minimum. If you don't think you'll ever have a need to save packet data for later use, a terminal is a fine alternative to a full-fledged computer. This is especially true if you have no other use for a computer. On the other hand, you'll be missing out on much of the fun and versatility of packet by not having a computer in the shack.

So Which Computer is Best?

If you're a little unsure about packet radio and you don't want to spend a lot of money, I highly recommend that you start with an inexpensive used computer. You'll find cheap Apples, Commodores, Color Computers, IBMs and other models at hamfest fleamarkets. If the computer is in good working order, it will be sufficient to get you started on packet. Make sure it has adequate *RAM* (random access memory) for the software you want to use. For small computers, 64 Kbytes or 128 Kbytes may be adequate. For larger machines—especially PCs—640 Kbytes is the safest bet. Also, skip any used computer that doesn't include at least one floppy disk drive.

If you have more ambitious plans for your computer, consider spending the extra dollars for the best *new* computer you can afford. High-speed microprocessors, expanded memory and VGA monitors are not cheap. However, you'll reap the benefits in the long run through the sheer pleasure you'll get from owning an advanced computer. As packet technology changes, you'll be able to change with it. In addition, you'll be able to run all the latest games and other software.

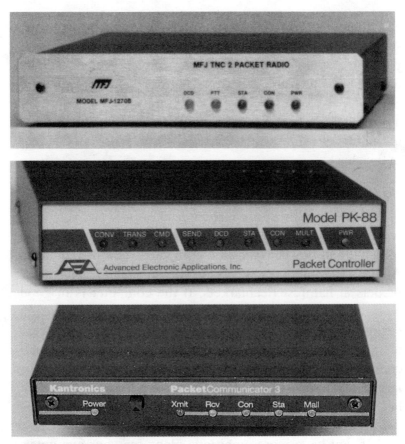

Here are just a few of the TNCs available. Some are basic while others offer innovative features such as "personal packet mailboxes." Shown from top to bottom: The MFJ 1270B, the AEA PK-88 and the Kantronics KPC-3.

The Terminal Node Controller (TNC)

The terminal node controller—or TNC—is at the heart of every packet station. As we discussed earlier, a TNC is a computer unto itself. It contains the AX.25 packet protocol software along with other enhancements depending on the manufacturer. The TNC communicates with you through your

If you want to operate other digital communications modes *in addition* to packet, multimode communications processors (MCPs) are ideal. Shown from top to bottom: The MFJ-1278, the AEA PK-232MBX and the Kantronics KAM Plus.

computer or data terminal. It also allows you to communicate with other hams by feeding packet data to your transceiver.

You have plenty of TNCs to choose from these days. The amount of money you'll spend depends directly on what you want to accomplish. If you're only interested in packet

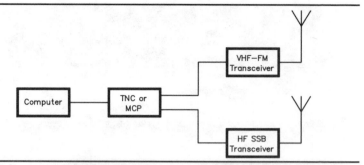

Fig 2-2—A typical packet station consists of a computer (running terminal emulation software), a terminal node controller (TNC) and a VHF-FM transceiver. If you want to be active on HF packet, an SSB HF-transceiver is required as well. If the cost of a computer is beyond your budget at the moment, consider a used data terminal instead.

operating, a basic TNC is all you need. Many of these TNCs include convenient features such as personal packet mailboxes where friends can leave messages when you're not at home.

If you're willing to spend more money, you can buy a complete multimode communications processor, or MCP. These devices not only offer packet, they also provide the capability to operate RTTY, CW, AMTOR, fax and other modes. In other words, an MCP gives you just about every digital mode in one box.

TNCs and MCPs must connect to your computer or data terminal *and* your transceiver (see Fig 2-2). It would sure make life easier if all connectors were standardized, but this is not the case. Fortunately, TNC and MCP manufacturers provide instructions for connecting their devices to your station. If you're lucky, you'll be able to use standard cables that can be purchased at just about any computer or electronics store. If not, you'll need to get out your soldering iron and assemble some custom cables of your own. Just follow the instructions and check your finished product carefully before attempting to use it.

Using Your Computer as a TNC

Throughout this chapter we've been discussing TNCs and MCPs as *outboard* devices separate from the computer itself. Does it always have to be this way?

Several years ago, a software package known as *DIGICOM>64* was developed by a group of German amateurs. DIGICOM>64 was designed to run on Commodore computers. It emulated most of the functions of a packet TNC *and* a terminal, requiring only a simple outboard modem to act as the interface to the transceiver. (The modem can be built for less than $30.) DIGICOM>64 was a boon to amateurs who were interested in packet, but who couldn't afford a full-featured TNC. Since its introduction, it has become very popular among Commodore users. While DIGICOM>64 can't compete with TNCs when it comes to features and flexibility, it is more than adequate for general packet operating. You'll find sources for DIGICOM software and modem kits listed in Table 2-1.

More recently, IBM PC and compatible users finally got their own TNC-emulation systems. One is known as *BayCom*. Like DIGICOM>64, BayCom uses the PC to emulate the functions of a TNC/terminal while a small external modem handles the interfacing. BayCom packages are available in kit form for roughly half the price of a basic TNC.

PC owners also have the option of buying full-featured TNCs that mount *inside* their computers. Several of these TNC *cards* are available from various manufacturers. They are complete TNCs that plug into card slots inside the computer cabinet. No TNC-to-computer cables are necessary. Connectors are provided for cables that attach to your

Not all TNCs and MCPs are stand-alone units. For example, DRSI-2000 (above) and the AEA PCB-88 (below) plug into card slots *inside* IBM-PCs and compatibles.

transceiver. In many cases, specialized software is also provided for extremely efficient operation.

The Radio: Your Link to the Packet World

When it comes to transceivers, packet isn't picky. I've communicated with packet using everything from modern rigs to 25-year-old tube transceivers. This is not to say that *any* radio can be used for packet. Some older HF rigs cannot switch fast enough from transmit to receive (and vice versa). Some VHF radios distort the packet tones as they are transmitted or received. Even so, these types of rigs are the exception, not the rule.

The audio output from the packet TNC is usually fed to the microphone input. Received audio from the rig is often obtained from an external speaker jack. If your radio doesn't have an external speaker jack, you may have to tap the audio directly at the speaker with a switch or a Y connector.

When connecting a TNC to a transceiver, there are several items to consider. . .

❑ *AFSK Audio*: Your TNC translates digital packet data into audio tones that shift rapidly from one frequency to another. This is known as *AFSK*, or audio frequency shift keying. The AFSK audio must be supplied to your transceiver so that packet data can be sent to other stations. The critical factor here is *audio level*. It's important to avoid overdriving or underdriving your microphone input amplifiers. Overdriving causes distortion. Since packet requires every bit of received data to be error-free, you must minimize distortion in your transmitted audio. On the other hand, underdriving the input will result in weak, anemic signals that will be difficult for the receiving station to decode. Listen to your packet transmissions on another receiver and compare your signal to others. Or, have a friend listen to your transmissions and offer his or her opinion. You'll find that all TNCs provide a means to adjust the input audio level—either with a potentiometer, a switch or a circuit board jumper. Finding the correct input audio level is not difficult since the TNC and the radio will usually give you plenty of latitude.

❑ *Receive Audio*: This is the audio that is fed to your TNC *from* your transceiver. You need to supply just enough audio for the TNC to work its decoding magic. Too much audio will swamp the TNC and result in very little magic at all! When operating HF packet, increase your AUDIO LEVEL— or VOLUME—control until the DCD (or RCV) indicator on your TNC begins to glow. Now *decrease* the level until it just stops glowing. This is your optimum setting, although you may have to adjust it as you operate. If you're operating with a VHF-FM rig, open your transceiver's SQUELCH control and slowly increase the VOLUME until the DCD (or RCV) LED on the TNC begins to glow. Move the control just a tiny bit beyond this point and you should be set. Now close your

Get Keyed Up!

TNCs and MCPs use solid-state switching for transmitter control—particularly when operating RTTY, AMTOR, ASCII and packet. (Some MCPs employ an internal relay for CW keying.) Solid-state switching is fast and efficient. It's perfect for modern transceivers, but it can cause problems when applied to older rigs (particularly tube radios).

Marrying today's technology to yesterday's equipment can be a challenge, but it's not impossible. One easy solution is to buy a small 12-V relay and wire it to the TNC or MCP as shown in Fig 2-3. The relay acts as an isolator between the interface and the rig. The TNC or MCP keys the relay which, in turn, keys the radio. More elegant solutions are possible using solid-state devices. See "Cheap and Easy Control-Signal Level Converters" by James Galm, WB8WTS, in the February 1990 issue of *QST*, pages 24-27.

Another frequent problem involves the use of hand-held transceivers

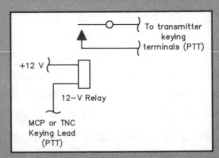

Fig 2-3—A small relay can be used to key older transceivers.

SQUELCH control until the LED stops glowing. If you want to work the weaker packet stations, leave your squelch "loose." If you're only concerned about strong local stations, increase your squelch considerably to block the weaker signals.

❑ *Keying*: The TNC must be able to switch your transceiver from transmit to receive automatically. Solid-state switching is commonly used because it's fast and efficient. Most transceivers manufactured since 1980 should be compatible with this type of switching. All that's required is a connection from the TNC to the PTT (*push-to-talk*) pin on the

for packet. Here are a few suggestions for keying ICOM, Kenwood and Yaesu hand-helds:

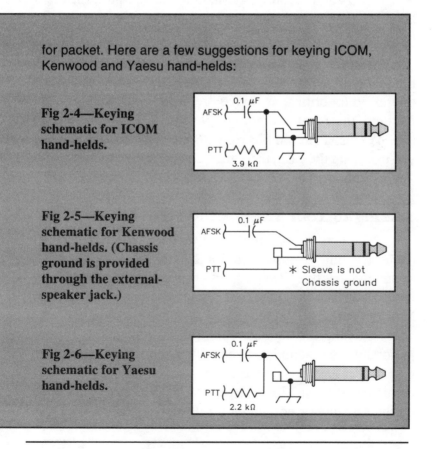

Fig 2-4—Keying schematic for ICOM hand-helds.

Fig 2-5—Keying schematic for Kenwood hand-helds. (Chassis ground is provided through the external-speaker jack.)

Fig 2-6—Keying schematic for Yaesu hand-helds.

While just about any transceiver can be used for VHF packet, rigs such as the Alinco DR-1200 are specifically designed for packet use.

microphone plug. (Consult your TNC manual.) Some older transceivers and certain hand-held models require different keying arrangements. See the sidebar "Get Keyed Up!" for more information.

Several manufacturers have created VHF-FM transceivers designed *specifically* for packet. These radios are optimized for critical audio requirements and contain special features that make them usable for high-speed packet applications. Your old FM rig will certainly do the job but, if you can spare the cash, these transceivers are definitely worth your attention.

Talking to Your TNC for the First Time

Once you've soldered every wire and connected every cable, it's time to turn on the juice and see what happens. Assuming that there are no sparks or smoke, the first priority is to get your TNC to communicate with your computer or data terminal.

Matching the data communications parameters is very important. For example, if your computer or data terminal is communicating at 4800 bits per second and the TNC is communicating at 1200 bits per second, you'll see gibberish on the screen—or nothing at all. Both devices are trying to talk to each other, but the information is getting lost in translation!

Read your software instructions (if you're using terminal emulation software) as well as your TNC manual. Some TNCs are shipped with preset (or switch-set) communications parameters and all you'll need to do is configure them to match your system. Other TNCs use a technique known as *autobaud*. This means that the TNC or MCP sends a test message repeatedly at various speeds when it is activated for the first time. When you see the message appear in plain English, the TNC is communicating at the speed setting of your computer or data terminal. An autobaud routine may look something like this:

zxcvdfn23asvzmnxcvenfv9a09fv/ktmn34tnza4
PRESS (*) TO SET BAUD RATE
qqqqeoprwpeowperowpoeirp2eoaca.,cmwpdo

In this example, the trick is to press the proper key *quickly* the moment you see the English text. When the computer and TNC have established communications, you'll probably see a sign-on message followed by *a command* prompt:

cmd:

The first thing to do is enter your call sign. (Throughout this book, the symbol <CR> represents the ENTER or RETURN key on your computer or terminal.) We'll do this with the MYCALL command. For example:

cmd: MYCALL WB8IMY <CR>

MYCALL is just one of literally *dozens* of TNC parameters you can modify to match your operating conditions. With the exception of MYCALL and a few others, most of the parameters are preset at the factory (called *defaults*). This does not mean you have to leave them that way, however. For example, if your transceiver requires a longer interval between the time it is keyed and the time the packet signal is sent, you can lengthen the TXDELAY setting. I found this to be the case when I used my TNC with an older FM transceiver. I had difficulty connecting to other stations until I performed some tests and discovered that my TNC was keying the rig and applying the AFSK audio too quickly. A slight adjustment of my TXDELAY solved the problem.

Before changing a parameter, it's always best to consult your manual. Some parameters are relatively harmless (such as BTEXT and CTEXT) and can be altered at will. Other parameters are very important and can cause a great deal of frustration if you tamper with them. Keep notes of whatever changes you make—the old settings and the new. You'll probably discover that the factory defaults are adequate to get you started. Once you have more experience with your system, you can begin tailoring your TNC to your particular operating habits.

Flow Control

As your TNC exchanges information with your computer or terminal, the data flows back and forth at a high rate of speed. Despite this efficiency, you'll sometimes send data faster than your TNC can handle. At other times, your TNC may send its data faster than your computer or terminal can tolerate. What we need is a means to control the back and forth flow of data in the same way that a traffic light controls the flow of traffic on a street.

Some computers and terminals achieve this through what is known as *hardware flow control*. Two separate wires (labeled CTS and RTS) on the EIA-232-E (RS-232) cable between the computer/terminal and the TNC are dedicated as control lines to start or stop the flow of data.

The most popular method, however, is *software flow*

control. With this method, special software characters are sent by the computer/terminal and the TNC to halt data flow temporarily. The tricky part of this scheme is that the TNC and the terminal software must *both* be using the same control characters. In most cases, the TNC and/or the terminal software will allow you to change the characters as necessary. Check the **XOFF** and **XON** parameters in your TNC to see which characters are in use. You can easily change them to match your software. Your TNC manual will provide more information on this important procedure.

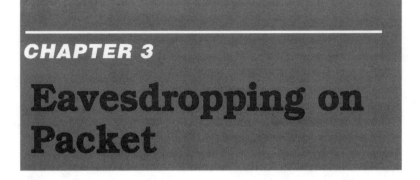

CHAPTER 3

Eavesdropping on Packet

hen I purchased my first TNC, I couldn't wait to hook it up and get on the air. I'd been listening to the local 2-meter packet frequencies for weeks. After a steady diet of screeches and buzzes, I was eager to learn the meaning behind those strange signals.

When I finished all the necessary wiring and programming, I switched to an active frequency. Now the moment of discovery was at hand. As I turned to my computer monitor I was thrilled to see . . . well, something altogether different than what I expected!

Excitement quickly melted into bewildered confusion. Why were there so many call signs? What did all those odd abbreviations mean? I could see readable text, but the same sentences kept appearing over and over. What was going on here?

Clearing Out the Cobwebs

One of my chief reasons for writing this book was to clear away the cobwebs of confusion that envelop so many new packeteers. If you've ever operated RTTY or AMTOR, you're probably used to plain English text flowing across your screen—and nothing else. Packet gives you text as well, but it also carries additional information with each transmission.

Sometimes this information is visible to you. Sometimes it isn't.

In this chapter we'll take a look at some of the typical packet transmissions you'll receive at your station. Our goal is to separate what's important from what isn't—and to explain some packet oddities along the way.

With all the different types of packet software on the market today, it's impossible to provide examples that will identically match what you'll see on your screen. Even so, you'll find more similarities than differences. Why? It all has to do with the amateur packet protocol we introduced in Chapter 1: AX.25. AX.25 isn't the only amateur packet protocol in existence, but it has the distinction of being the most widely used. Any software that's designed for use on AX.25-based packet networks *must* conform to the AX.25 protocols. You could think of it in terms of spoken languages. English is spoken with many accents, but we understand each other fairly well. That's because we all follow the same grammatical rules (more or less!) that govern English as a language.

Monitoring a Packet QSO

Live, or *real time* packet conversations are very enjoyable. They're common on the HF bands; a little less common on VHF—especially in crowded areas. Let's assume that you've found a couple of hams having an informal chat . . .

KR1S>KE3Z
Do you think the Patriots have a chance at the Super Bowl? I've been watching every game so far this season.
>>

KE3Z>KR1S
Nah. I've told you before, Jim, they're like the Red Sox with helmets. Know what I mean? >>

KR1S>KE3Z
What about their offense? They looked pretty good against the Giants last week. >>

KE3Z>KR1S
Yeah, but who won? The Giants! >>

Presuming you're able to receive signals from both stations, this is what you're likely to see on your screen. The > between each set of call signs indicates who is transmitting

---

Packet Tips: Who's on the Frequency?

You'd like to see which stations are using a particular packet frequency, but you don't have the time to sit down and monitor the activity yourself. No problem! Your TNC will tell you who's been on the frequency and *when* they were on.

First make sure the internal clock/calendar in your TNC is set properly. On most TNCs you'll use the DAYTIME command (or just "DA") to set it. The format is yearmonthdayhourminutesecond (only the last two digits of the year are used). There are no spaces, colons or slashes. For example, **DA 920601143030,** means 1992, June 1, 14 hours, 30 minutes and 30 seconds.

Your TNC keeps a constant log of every station it hears (up to a limit of about 20, depending on the TNC). Whenever you want to see this log, simply enter **MH** (for *Mheard*) at the **cmd** (command) prompt. All the station call signs will be displayed along with the date and time they were last heard—*if* you remembered to set your DAYTIME command properly, that is! An asterisk will appear beside call signs of stations that were heard through nodes or digipeaters. If you want to erase the log, simply enter **MHC** at the **cmd** prompt.

If you're curious to see who's using a particular frequency, park your receiver there and let your TNC do the listening!

---

and who is receiving. **KE3Z>KR1S** indicates that KE3Z is sending to KR1S.

Can you guess the meaning of the **>>** characters at the end of each comment and response? They're the packet equivalent of "over to you," or, "it's your turn to speak." Some packeteers use other symbols such as the Morse code "K" to accomplish the same purpose. Watch what happens to the flow of the same conversation when the symbols are omitted.

KR1S>KE3Z
Do you think the Patriots have a chance at the Super Bowl?

KE3Z>KR1S
Nah. I've told you before, Jim, they're like the Red Sox

KR1S>KE3Z
Like the Red Sox? How do you make that comparison?

KE3Z>KR1S
with helmets. Know what I mean?

KR1S>KE3Z
I wasn't talking about helmets. What about offense? They looked good

KE3Z>KR1S
Compared to who?

KR1S>KE3Z
against the Giants last week.

What a confusing mess! This is what happens when one station doesn't know when another station has sent a complete message. If one or both hams are slow typists—or if interference on the frequency is causing the TNCs to send several repeats of the same packets—it's easy to be fooled into thinking that the end of a sentence is the end of a whole message.

Enter the Node

Let's add a new twist to our sample conversation. What if the stations can't send packets to each other directly? They'll have to use *nodes* or *digipeaters* to bridge the distance and relay packets from one station to the other. (We'll discuss these interesting devices in Chapter 4.) If a node or digipeater is involved in the path, here's what you might see . . .

KR1S>KE3Z,N1BKE-5*
Do you think the Patriots have a chance at the Super Bowl?
I've been watching every game so far this season. >>

KE3Z>KR1S,N1BKE-5*
Nah. I've told you before, Jim, they're like the Red Sox with helmets. Know what I mean? >>

KR1S>KE3Z,N1BKE-5*
What about their offense? They looked pretty good against the Giants last week. >>

KE3Z>KR1S,N1BKE-5*
Yeah, but who won? The Giants! >>

Notice that N1BKE-5 has been added to the call sign list. In this case, N1BKE-5 is the call sign of a local packet node. Two questions come to mind immediately: why is there a **-5** after N1BKE's call sign and why does it include an asterisk as well?

The -5 is known as an SSID, or *secondary station identifier*. AX.25 packet protocol uses the SSID to distinguish between different station functions. N1BKE-5 is clearly a packet node, but what if N1BKE operates a packet bulletin board, too? How would you connect to it and not the node? Easy! N1BKE could designate his PBBS as N1BKE-*4*. Now you can connect to N1BKE-5 if you want to use the node, or N1BKE-4 if you want to check in to the bulletin board.

The SSID can be any number from 0 to 15. (An SSID of

0 is not displayed.) If you do not specify an SSID when you enter your call sign into your TNC, it assumes your SSID is zero. Most hams do not include an SSID in their MYCALL parameter unless they are using the TNC for a special purpose, such as a node or PBBS separate from their regular stations.

But what about that asterisk? The asterisk is used to let you know you're receiving packets that have been relayed through a node or digipeater. In the example above, you were not receiving signals from KE3Z or KR1S directly. Instead, you monitored their conversation as it was relayed through the N1BKE-5 node. Your TNC is trying to make it clear that you're copying the transmissions of N1BKE-5, not the other stations.

Repeat Transmissions. . . Repeat Transmissions. . .

Let's make our sample conversation a bit more complicated. What do you think you'd see if you were able to receive the transmissions of KE3Z *in addition* to those from the N1BKE-5 node?

KR1S>KE3Z,N1BKE-5*
Do you think the Patriots have a chance at the Super Bowl? I've been watching every game so far this season. >>

KE3Z>KR1S,N1BKE-5
Nah. I've told you before, Jim, they're like the Red Sox with helmets. Know what I mean? >>

KE3Z>KR1S,N1BKE-5*
Nah. I've told you before, Jim, they're like the Red Sox with helmets. Know what I mean?>>

KR1S>KE3Z,N1BKE-5*
What about their offense? They looked pretty good against the Giants last week. >>

KE3Z>KR1S,N1BKE-5
Yeah, but who won? The Giants! >>

KE3Z>KR1S,N1BKE-5*
Yeah, but who won? The Giants! >>

You're seeing each one of KE3Z's transmissions *twice*—once directly from his station and again through the N1BKE-5 node (note the asterisk that follows N1BKE-5 in the second transmission). Repeating messages can also appear if one station is having difficulty receiving another. As you may recall from Chapter 1, if the originating station does not receive an acknowledgment (ACK) from the receiving station, the packet is retransmitted. Let's watch the conversation again (without the node), but this time we'll introduce less-than-ideal conditions.

KR1S>KE3Z
Do you think the Patriots have a chance at the Super Bowl? I've been watching every game so far this season. >>

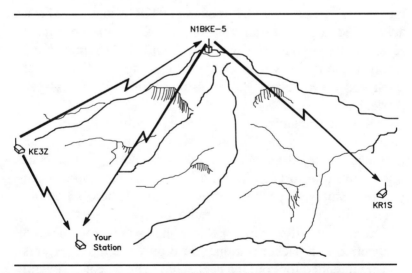

Fig 3-1—If you were monitoring an exchange between KE3Z and KR1S, you may see KE3Z's transmissions *twice*: Direct from KE3Z as well as from the N1BKE-5 node.

KE3Z>KR1S
Nah. I've told you before, Jim, they're like the Red Sox
with helmets. Know what I mean? >>

KE3Z>KR1S
Nah. I've told you before, Jim, they're like the Red Sox
with helmets. Know what I mean? >>

KE3Z>KR1S
Nah. I've told you before, Jim, they're like the Red Sox
with helmets. Know what I mean? >>

KR1S>KE3Z
What about their offense? They looked pretty good
against the Giants last week. >>

KE3Z>KR1S Yeah, but who won? The Giants! >>

KE3Z's packets are repeated several times because his
station did not receive prompt acknowledgments from KR1S.
This may have been caused by weak signal conditions or
interference on *either* end. For example, KR1S may have been
acknowledging each transmission, but KE3Z may not have
received the ACKs. On the other hand, KR1S may have
received the packets with errors, causing the TNC to reject
them and await repeat transmissions (otherwise known as
retries).

Repeat transmission can be very confusing for new
packeteers. If several different stations are active on the
frequency, repeat transmissions can quickly fill your screen!
The main point to keep in mind is that this is *normal* when
you're monitoring packet communications. There's nothing
wrong with your TNC or your radio.

It may comfort you to know that when you make a
connection to a packet station, you'll only see those packets
that are intended for you. If a packet has to be repeated several
times before you receive it error-free, you'll only see text
when it finally makes it through unscathed.

Packet Tips: Watching the Weak Ones

Under normal circumstances, your TNC will only display those packets it receives without errors. This is fine most of the time, but what if you hear some weak packet signals that aren't strong enough to be received error-free? The S-meter on your transceiver says there's a signal on the frequency, but your screen remains blank. Is there any way to copy those weak signals?

If you don't mind seeing errors in the text, the answer is *yes*. At the **cmd** prompt, enter **PASSALL ON**. Now your TNC will display packets even if they contain errors. You'll see some unusual characters on your screen, but your TNC may decode enough data to make the text readable. With any luck you'll be able to find out where those weak signals are coming from. This is a handy feature to use when you're exploring packet activity in your area. In most TNCs, your MHEARD logging is disabled whenever the PASSALL is on.

There *is* a TNC parameter that will allow you to continue to monitor other packets even when you're connected to another station. It's called *MCON*. Why would you want your screen cluttered with other packets when you're trying to communicate with someone else? Well, there are a few situations where this ability comes in handy—as we'll see in later chapters.

Watching the Bulletin Boards

Packet bulletin boards (PBBSs) often function as nerve centers for amateur packet networks. PBBSs are collectors and distributors of information. By connecting to your local PBBS, you can send mail, receive mail, send bulletins, read bulletins, upload files, download files and so on. Many packeteers enjoy connecting to local PBBSs every couple of days to read new bulletins. This type of activity makes PBBSs easy to spot.

As you monitor the packet frequencies, look for transmissions similar to the one shown below.

```
WB8SVN-4>WB8ITK
MSG # TR   SIZE TO        FROM   @BBS   DATE     TITLE
7287 B$    2016 QST       W1AW   ARL    920427   ARLB037 Propagation
7286 B#    2409 ALL       WP4IIW USBBS  920427   Alinco + Kenwood
7284 B#    716 WANTED W1KSZ NEBBS 920427   HF TRANSCEIVER
```

If the PBBS has to send this packet more than once before it receives an acknowledgment from WB8ITK, you'll see. . .

```
WB8SVN-4>WB8ITK
MSG # TR   SIZE TO        FROM   @BBS   DATE     TITLE
7287 B$    2016 QST       W1AW   ARL    920427   ARLB037 Propagation
7286 B#    2409 ALL       WP4IIW USBBS  920427   Alinco + Kenwood
7284 B#    716 WANTED W1KSZ NEBBS 920427   HF TRANSCEIVER

WB8SVN-4>WB8ITK
MSG # TR   SIZE TO        FROM   @BBS   DATE     TITLE
7287 B$    2016 QST       W1AW   ARL    920427   ARLB037 Propagation
7286 B#    2409 ALL       WP4IIW USBBS  920427   Alinco + Kenwood
7284 B#    716 WANTED W1KSZ NEBBS 920427   HF TRANSCEIVER

WB8SVN-4>WB8ITK
MSG # TR   SIZE TO        FROM   @BBS   DATE     TITLE
7287 B$    2016 QST       W1AW   ARL    920427   ARLB037 Propagation
7286 B#    2409 ALL       WP4IIW USBBS  920427   Alinco + Kenwood
7284 B#    716 WANTED W1KSZ NEBBS 920427   HF TRANSCEIVER
```

Since you'll be making use of PBBSs yourself, it's a good idea to note the complete call signs of all systems you find. Make a note of their frequencies as well.

When a PBBS is not connected to other packet stations, it will occasionally transmit identifications. PBBSs on the VHF and UHF bands frequently use nodes or digipeaters to relay their IDs over a wide area.

KY1T-4>MAIL,KB1CQ-7,KY1T-2*,KF1ET
Mail for: KR1S, KE3Z, AA2Z, N1BKE

In this case, you've picked up the ID transmission of the KY1T-4 PBBS. The ID was relayed through the KB1CQ-7, KY1T-2 and KF1ET nodes. Did you notice the asterisk immediately following KY1T-2? This means that you did not

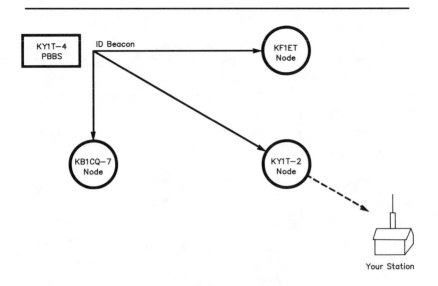

Fig 3-2—When the KY1T-4 PBBS sends its ID beacon, it may use several nodes to relay it over a wide area. In this case, you receive the ID that was relayed via the KY1T-2 node.

receive this transmission directly from the PBBS. Instead, you received a relay from KY1T-2.

The KY1T-4 PBBS is announcing its availability to the local network and it's displaying a list of hams who have mail waiting. If your call sign appears on the list, you have a message waiting for you on the PBBS!

It Looks Like a PBBS, But . . .

Many packet TNCs and MCPs incorporate *personal packet mailboxes* in their designs. They may look like PBBSs when you see them on the air, but there are major differences.

[KPC2-5.00-HM$]
16000 BYTES AVAILABLE
Thanks for checking in. Please leave a message and I'll get back to you. 73, Bill/W1KKF.

ENTER COMMAND: B,J,K,L,R,S, or Help >

Notice, for example, that this packet mailbox has only 16,000 bytes of data storage available. Packet mailboxes store all of their incoming and outgoing messages in solid-state RAM (random access memory). The typical RAM capacity of a packet mailbox is only about 32 Kbytes. True PBBSs, on the other hand, usually store data on *hard disks* with several *megabytes* of capacity.

You'll also notice that the *command line* (**ENTER COMMAND: B,J,K,L,R,S, or Help >**) is much shorter than those you'll see on most packet bulletin boards. That's because packet mailboxes have very limited capabilities. This one will allow you only to send mail (S), read mail (R), list messages (L), kill messages (K), see a list of stations heard recently (J) and disconnect (B).

With their meager memories and limited features, packet mailboxes are not intended to perform like real PBBSs. Instead, packet mailboxes are convenience devices. They allow stations to connect and exchange mail—even when the control operators are not present. Some hams leave their mailboxes on around the clock. This makes it easy for friends to drop off messages (and read replies) any time of the day or night.

If your packet mailbox uses the proper software, it's even possible to arrange *automatic mail forwarding* to and from your local PBBS. If you have mail waiting for you at the PBBS, it will connect to your mailbox and deliver the messages automatically. You can enter an outgoing message in your mailbox and your TNC or MCP will transfer it to the PBBS. This feature is particularly handy when you need to post a message on the PBBS, but the system is busy. Just leave it in your mailbox and let your TNC/MCP worry about handing it off to the bulletin board!

Packet mailboxes are very popular on the VHF and UHF bands. However, it's illegal to operate a packet mailbox on the

HF bands without a control operator present.

While you're eavesdropping on the VHF and UHF bands, it's likely that you'll find many hams with personal packet mailboxes. It's worthwhile to note their call signs and frequencies, too. You never know when you'll need to send a message to someone at 4 in the morning!

Watching the Data Flow

When a packet bulletin board isn't busy handling requests from users, it's usually sending or receiving data from other PBBSs. Most modern packet networks conduct their data exchanges on other bands. A PBBS on 2 meters, for example, usually sends and receives its messages on a 222- or 420-MHz *backbone* link.

Some of these backbone networks operate at data rates beyond 1200 bits per second. Regardless of the data rate, backbone links are strictly off-limits to individual users. If you attempt to connect to a backbone, you'll receive an abrupt *disconnect* response!

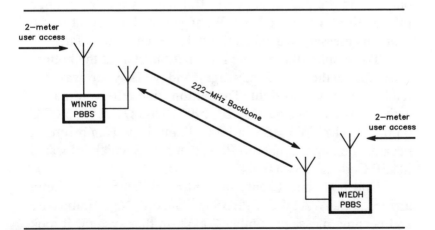

Fig 3-3—Most PBBSs have a b*ackbone* link that they use to forward messages from one system to another. Backbone links are often separate from frequencies used for normal access.

A number of PBBSs also have the capability to relay data on HF frequencies. In fact, HF mail forwarding is an essential part of the global packet network.

Whether the forwarding is taking place on UHF, VHF or HF, the text on your screen will look pretty much the same. What you're seeing is a "conversation" taking place between two packet bulletin board systems.

W1EDH-7>W1NRG-9
R:920527/1757z @:W1EDH.CT.USA.NA Glastonbury, CT #:27008 Z:06033
R:920527/1751z @:WA1TPP.MA.USA.NA Granville, MA #:4431 Z:01034
R:920527/1733 @:WB1DSW.NH.USA.NA E. Kingston, NH #:52319 Z:03827
R:920527/0644 @:WA1WOK.NH.USA.NA Concord #:20062 Z:03301
R:920527/0515 @:WA1YTW.NH.USA.NA Fitzwilliam #:8719 Z:03447

I am looking for the QSL address for 6Y5EF in Jamaica. It's only good in a 1992 Callbook. Can anyone help me? Tnx de Bruce KA1TWX @ WA1YTW.NH.NA

The text of the message looks clear enough, doesn't it? This fellow needs assistance and he has decided to place a bulletin on the packet network. But what is all that strange information above the text? What you're looking at is the *route* his message has taken through the network so far.

To decipher the routing information, start at the bottom (just above the actual message) and work your way up, reading from left to right. The bottom line tells you that he entered his message on May 27, 1992 (920527), at 0515 UTC on the WA1YTW PBBS in Fitzwilliam, New Hampshire. It became message number 8719 on that PBBS which is located in ZIP Code (Z:) area 03447.

One hour and twenty-nine minutes later, his bulletin arrived at the WA1WOK PBBS in Concord, New Hampshire and became message number 20062 on that system. It took almost 11 hours to finally reach the WB1DSW system in East Kingston, New Hampshire at 1733 UTC. A problem in the backbone system may have caused the delay. (Remember

Murphy's Law: If anything can go wrong, it will go wrong!) In less than 20 minutes the bulletin had reached the WA1TPP PBBS in Granville, Massachusetts and was passed along to W1EDH in Glastonbury, Connecticut six minutes later at 1757 UTC. At the very top of the routing information you can see that W1EDH-7 is in the process of forwarding a copy of the bulletin to W1NRG-9. The W1NRG PBBS will eventually pass it along as well.

When you read a message on a packet bulletin board, you won't normally see such detailed routing information. However, it *is* available and you can see it if you wish. (Ask your SysOp which command to use to read the routing information.) The ability to understand routing comes in handy when you're trying to determine why a piece of mail arrived late. It's also fascinating to examine the paths messages take from their points of origin to your local PBBS.

DX PacketClusters

Another interesting phenomenon you're likely to see, particularly on VHF, is a DX *PacketCluster*. A *PacketCluster* is a network of specialized nodes developed for DX hunting and contesting. *PacketCluster* nodes allow many stations to connect simultaneously and exchange DX information. When a station connects to the cluster, you'll see a response like this:

KC8PE>WB8IMY
Hi Steve! Welcome to YCCC Packetcluster node - Cheshire CT
Cluster: 47 nodes, 8 local / 490 total users Max users 609 Uptime 23:17
WB8IMY de KC8PE   27-Apr-1992 2347Z   Type H or ? for help >

In most cases, the first thing a user does is request a list of the five most recent DX sightings (often referred to as *spots*). Assuming that you can receive transmissions from the station as well as the *PacketCluster* node, this is the exchange you're likely to receive:

WB8IMY>KC8PE
SHOW/DX

```
KC8PE>WB8IMY
14081.0   SV9/SV1AHH 27-Apr-1992  2346Z   RTTY CQ   <WA2IZN>
14195.1   CY0SAB      27-Apr-1992  2341Z             <VE2FN>
14190.0   4S1GHE      27-Apr-1992  2344Z             <WB2IKL>
 7017.2   CE3ZW       27-Apr-1992  2343Z             <W1YY>
 7004.2   ZA1TAJ      27-Apr-1992  2342Z             <W1YY>
WB8IMY de KC8PE       27-Apr 2347Z >
```

In this example, WB8IMY requested the list by sending the **SHOW/DX** command. KC8PE, the *PacketCluster* node, responded with a list of the five most recent spots. The list shows the frequencies, DX call signs, dates, times (UTC) and the call signs of the stations who posted the information on the network.

It's common for a station to remain connected to a DX *PacketCluster* while searching for new countries to contact. When a new DX station is spotted and added to the cluster, the information is sent to *all* connected stations. If you're monitoring a *PacketCluster* frequency, you'll see these distribution transmissions taking place.

KC8PE>WB8IMY
DX de K2PBP: **28461.9**   **V73CT**   **0021Z**

KC8PE>NF1J
DX de K2PBP: **28461.9**   **V73CT**   **0021Z**

KC8PE>KR1S
DX de K2PBP: **28461.9**   **V73CT**   **0021Z**

KC8PE>WA1TRY
DX de K2PBP: **28461.9**   **V73CT**   **0021Z**

The KC8PE node is letting everyone know that K2PBP has found V73CT on 28.461.9 MHz. Every ham connected to the cluster who needs V73CT will be stampeding to the 10-meter band right away! DX *PacketClusters* offer many other features and we'll explore them in greater detail in Chapter 4.

When Packet Really Gets Busy!

So what's it like to sit back and monitor the activity on a

Packet Tips: "Invisible" Information

When you're monitoring a packet frequency, are you really seeing *everything* that's being transmitted? No, you're not. In addition to all the text and other data, there are *control packets* flying back and forth between stations. Control packets contain connect requests, disconnect requests, disconnect modes (when a station is "busy" and unable to accept additional connections) and so on.

You can see these control packets by turning your **MCOM** parameter **ON**. (Some TNCs don't use the MCOM command. Instead, it's a variation of the MONITOR command. Check your manual.) This function is especially useful when operating HF packet. For example, let's say that I've tuned in to the transmissions of N6ATQ as he's attempting to establish a connection to NF9T. Normally, I'd hear his transmissions, but nothing would print on my screen. By activating my MCOM function, I'd see the following:

N6ATQ>NF9T <<C>>
N6ATQ>NF9T <<C>>
N6ATQ>NF9T <<C>>
N6ATQ>NF9T <<C>>
N6ATQ>NF9T <<C>>
N6ATQ>NF9T <<C>>
N6ATQ>NF9T <<C>>
N6ATQ>NF9T <<C>>
N6ATQ>NF9T <<C>>
N6ATQ>NF9T <<C>>

N6ATQ has just sent 10 connect requests (<<C>>) to NF9T without getting a response. It's likely that N6ATQ's TNC will stop trying after the 10th attempt, giving me an opportunity to connect to him. Without using my MCOM function, I wouldn't have even known he was there!

busy packet frequency? Now that you've read this chapter, I hope you'll be able to make sense out of what may seem like pure chaos at times. Here's a brief glimpse of the packet action on 20 meters on a weekday evening. There are several stations on the frequency, but packet manages to share the limited space fairly well. . .

K4EID>K4TKU:
It didn't look all that bad inside, but it has
layered circuit boards and I didn't really want to

K4EID>K4TKU:
tear into it. I'll send it back for an estimate. >>

KT4NC>WD4OXT-14:
###RETRIED OUT AT NODE K4GHR

XF3R>XE1M-15:
Another ham (80 years old!) told me about a new mode called PACTOR.

XF3R>XE1M-15:
Another ham (80 years old!) told me about a new mode called PACTOR.

XF3R>XE1M-15:
Another ham (80 years old!) told me about a new mode called PACTOR.

XF3R>XE1M-15:
I never heard of this before!

XF3R>XE1M-15:
Has anybody further info ?

XF3R>XE1M-15:
Thanx in advance.

XF3R>XE1M-15:
73...Joss.

Can you sort out the activity that's taking place? First of all, it looks like K4EID is carrying on a conversation with K4TKU. The KT4NC node quickly squeezes in a transmission to let WD4OXT-14 know that his attempted connection to the K4GHR node was unsuccessful. Finally, the XF3R PBBS makes a number of transmissions as it forwards a bulletin to XE1M-15. (Did you notice the retries?)

It's a good idea to monitor the packet action before attempting to get on the air yourself. Take plenty of time to find out what's going on in your area. When you feel confident enough to make your first packet connection, you'll be equipped with the knowledge you've gained from watching everyone else. The result will be a more enjoyable (and less confusing!) experience.

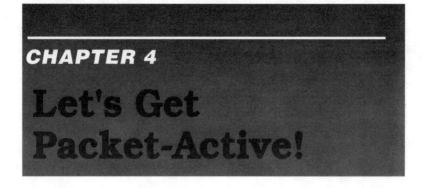

CHAPTER 4

Let's Get Packet-Active!

y now it's safe to assume that you've spent some time monitoring packet activity. It's fascinating to watch packet in action, but the novelty wears off when you're just a spectator and not a participant. Your packet station is set up and ready to go, isn't it? Good! It's time to start using the experience you've gained so far. Let's begin with some *real time* (keyboard-to-keyboard) packet conversations.

Your First VHF Connection

If you have a friend who's active on packet, why not make an appointment to meet at a certain frequency and time? That's probably the easiest way to enjoy your first packet conversation.

During my first packet connection, I was fortunate to have a separate 2-meter transceiver that I used to maintain simultaneous voice contact with the other station. It helped iron-out a few wrinkles in my packet equipment! For example, my friend was having difficulty decoding my packets. After listening to several transmissions, he was able to tell that I had the audio output level of my TNC set too high (I was over-driving the microphone input of my FM transceiver).

"Steve, your audio is loud and distorted on my end. Back off the audio level control on your TNC and let's see what happens." Sure enough, all it took was a slight adjustment and my packets began to appear on his screen—nervous typing errors and all!

If no one is able to meet you on the air, go it alone. Park your transceiver on a packet frequency (see Table 4-1) and wait for your chance to connect to someone. Whether you call

Table 4-1

Popular Packet Frequencies

VHF

6 meters

50.60 through 51.78 MHz
50.62 is the 6-meter packet calling frequency

2 meters

Every 20 kHz from 144.91 through 145.09 MHz.
Packet can also be found between 145.50 and 145.80 MHz.

222 MHz and Up

Packet can be found on the 222, 420, 902 and 1240-MHz bands, but bulletin boards and live QSOs are sporadic. Most of the activity on the higher bands is in the form of *backbone* links that pass traffic between bulletin boards and nodes. Avoid these backbones; they're not intended for individual user access.

HF
3606 kHz
3630 kHz
3642 kHz
7093 kHz
7097 kHz
10145 kHz
14101-14105 kHz
18100-18110 kHz
21099-21105 kHz
28099-28105 kHz

a friend or a total stranger, the format is always the same: you must send a *connect request*.

Place your TNC in the *command mode* if it isn't there already. Whenever you're connecting directly from your station to another station, you must be in the command mode. With most types of terminal software, you'll see a **cmd:** prompt on your screen. The exception is software that operates in what is known as the *host* mode. Host software streamlines TNC-to-computer communication and makes it more user-friendly. With host software, the **cmd** prompt may not appear.

For the moment, let's assume that your TNC is running in the standard terminal interface mode. You notice that WA1TRY is using the local PBBS, but you just saw him transmit a "B" (for "Bye"). He's about to disconnect and it's your time to pounce! Enter

CONNECT WA1TRY or just **C WA1TRY**.

Watch your transceiver and your TNC. As soon as the frequency is clear, the TNC will key your rig and send its first connect request. If the path between you and WA1TRY is good, you'll see the following message very quickly:

*** CONNECTED to WA1TRY

You've done it! Your station is now connected to WA1TRY. Your TNC will switch automatically to the *converse* mode. This means that everything you type will be sent directly to WA1TRY. The AX.25 error-checking protocols are active and they'll ensure that only error-free packets appear on your screen and his.

WA1TRY is probably wondering who connected to him and why. Send a short greeting:

Hi! My name is Steve and I live in Wallingford >>

Be patient as you wait for his response. Some hams are very slow typists!

Packet Tips: Calling CQ

If you can't find anyone to talk to, you may need to send a *CQ* (general call) to flush other hams out of the woodwork! You can send your CQ as a single, *unconnected* packet, but you must be in the converse mode to do so. (If you're in the command mode, enter **CONV** at the **cmd:** prompt.) When you're not connected to another station, everything you enter while in the converse mode is sent as an unconnected packet. Any station that receives the packet without errors will be able to read it. Since it's sent without an established connection, an ACK from another station isn't required or expected.

Sending a CQ this way is easy. Just type a brief message and hit your ENTER or RETURN key. For example:

CQ, CQ from NØMZR in Palmyra, Missouri. Anyone around?

The problem with this approach is that the CQ is sent only once. What if a burst of interference prevents someone from receiving it? What if you sent your CQ just seconds before a lonely packeteer arrived on the frequency? To send CQ more than once, you'd have to type the entire line over and over.

Fortunately, there's an easier way to send multiple CQs. Your TNC is equipped with a **BEACON** function. When the beacon is activated, your TNC will transmit unconnected packets repeatedly until you turn it off.

The content of your beacon packet is determined by your **BTEXT** parameter. BTEXT stands for "beacon text"

Hello, Steve. My name is Rich and I'm located in Meriden. I don't recognize your call sign. Are you new to packet? >>

And the conversation flows from there! If you don't know what to say, describe your station briefly, or ask what the other person does for a living. If you have questions about packet and it contains the actual text of your CQ message. Simply switch to the command mode and enter something like this:

BTEXT CQ CQ from N6ATQ in Escondido, California

When you're ready to start, enter the **BEACON** command along with the beacon *rate*.

BEACON EVERY N

N usually equals a certain interval of time. In some TNCs, for example, 1 is equal to 10 seconds. So, **BEACON EVERY 1** will cause the TNC to transmit an unconnected packet containing your BTEXT message every 10 seconds. **BEACON EVERY 3**, on the other hand, would cause the packet to be transmitted every 30 seconds (10 × 3). Timing can differ from one TNC to another, so check your manual carefully.

What beacon rate should you select? If the frequency is busy, keep your beacon rate low (once per minute or longer). If the frequency is unoccupied, you can beacon at a much faster rate (every 10 seconds). The key is knowing when to stop. Sending beacons on a quiet frequency should net a response within a minute or two. If it doesn't, turn off the beacon (**BEACON EVERY 0**) and try another frequency. If someone answers your call, you may need to switch to the command mode and turn off the beacon. Some TNCs will continue beaconing even after you've started a conversation with another station!

operating, don't be afraid to ask. Experienced packeteers are always glad to assist newcomers. Remember to use >> or some other easily recognized symbol to let other stations know when it's their turn to respond.

Most TNCs have an indicator on the front panel labeled STA. This stands for "status." Watch this indicator during your packet conversations. It'll give you a pretty good idea of the path quality between you and the other station. When you type a line of text and hit your ENTER or RETURN key, the STA indicator will glow immediately. This means that your TNC is attempting to deliver a packet. As soon as its error-free arrival is verified, the STA indicator will go off. If you're a fast typist and you're entering a long message, the STA indicator will glow continuously as the packets "stack up" in the TNC, each awaiting transmission and acknowledgment. When the last packet arrives safely at its destination, the STA indicator will finally switch off.

When the conversation ends, it's time to *disconnect*. The question of which station terminates the link depends on who has the last word... so to speak.

Thanks for the chat, Rich. I've got to grab a bite to eat. So long! >>

Now the ball is in Rich's court. He can fire off a final comment, or just terminate the link. When he breaks the link, you'll see ***** DISCONNECTED** on your screen. Unless you're sure that Rich has nothing more to say, it's rude to send "goodbye" and immediately break the link. Make sure the other station is really finished before you disconnect!

To terminate any packet link, all you have to do is return to the *command* mode and enter **D** or **DISCONNECT**.

Getting Connected on HF

If you're hunting for long-distance packet contacts, you'll find plenty of action on the HF bands! To explore HF packet, you'll need a stable SSB transceiver. Stability is important because HF packet is very unforgiving when it comes to drifting signals. If you own a tube-type radio, allow plenty of warm-up time before attempting to use it on packet.

An accurate tuning indicator is another valuable asset. Without a tuning indicator, it's difficult to know when you've tuned the signal correctly. An HF packet signal may sound

Packet Tips: Honoring the Band Plans

As you explore packet, you'll notice that activity seems to take place between certain groups of frequencies. The frequencies set aside for HF and VHF packet are determined not only by FCC regulations, but also by carefully designed *band plans*. Who creates these band plans? Your fellow amateurs!

You say you can't recall ever attending a band-planning meeting? Well, the meetings only take place when a new plan is created or an old plan is revised. Many of the discussions are informal in the early stages of the process. Representatives of Amateur Radio groups will exchange ideas and attempt to arrive at a general agreement. When a band plan is finally considered for endorsement by the ARRL, the details are published in *QST* and everyone is invited to comment.

Band plans are necessary to guarantee that all amateurs have enough spectrum to carry out their favorite activities. With the increasing popularity of packet, some packet subbands are becoming crowded. When you're trying to enjoy a packet conversation on a congested frequency, it's tempting to simply move elsewhere in the band. By violating the band plan, however, you may be causing tremendous interference to other amateurs.

HF and VHF packet transmissions are "wide" when it comes to bandwidth. On HF, a packet signal can wreck several CW, RTTY or AMTOR conversations at once. On VHF, packet can be devastating when operated in the *weak signal* (SSB/CW) or satellite portions of the band.

By honoring the band plans, you'll be respecting the agreements that bind us together as Amateur Radio operators. Designated HF and VHF packet frequencies are shown in Table 4-1.

loud and clear, but if you're off frequency by even a hundred hertz, your screen may remain blank. All MCPs and many TNCs include a tuning indicator. If not, it may be available as optional equipment.

Table 4-1 will give you an idea of the best frequencies to check for packet activity. Before you begin hunting, make sure your TNC or MCP is set up for HF packet. Some units require you to switch on the HF function and set the proper data rate. 300 bits per second is common on all HF bands, although 1200 bit/s is legal above 28 MHz.

Choose your band and tune *very* slowly through the packet signals. If you find a strong signal, watch your tuning indicator and carefully adjust your receive frequency until you're exactly on target. This may take some practice.

If a station is sending *information* packets (packets containing text), you should see the transmissions very shortly. On the other hand, if a station is trying to connect or disconnect, your screen will remain blank. To remedy this problem, go to the command mode and activate your MCOM or MONITOR function (depending on which TNC you're using). Now you'll see everything.

If you see a series of disconnects, you're in luck! You've discovered a station that's just finishing a connection with someone else. Stumbling upon a station sending connect requests isn't bad news either. Stay on the frequency and watch what happens. If the requests go unanswered, give him a call!

Before you attempt an HF packet connection, check your receive audio level. Adjust the level control until the RCV or DCD indicator on your TNC glows when a packet is received. If the indicator blinks in response to noise on the frequency, you have the audio level set too high.

HF and VHF packet are identical when it comes to making a connection. Just switch to the command mode and enter your connect request:

cmd: CONNECT LA4JBX

If the other station acknowledges your request, you'll be rewarded with:

*** CONNECTED to LA4JBX

The rest is up to you and the other station. Don't be surprised if the flow of the conversation is much slower on HF than VHF. Not only is the data rate much lower (300 vs 1200 bits per second), noise, fading and interference will often require both stations to retransmit their packets several times. Just because you don't see anything for a minute or two, it doesn't necessarily mean the connection is failing.

Trouble on the Frequency

If there's a problem with your connection to another station (HF or VHF), your STA indicator may provide the first clue. When the path is very good, all packets transmitted by your TNC will be received and acknowledged right away. You'll see the STA indicator flashing on and off as this takes place. If the STA remains on—even after you've sent a very brief message—it means that your TNC is still waiting for ACK signals from the other station. An STA that's still glowing long after you've entered the last line of your message is a sure sign of trouble. The culprit could be weak signals at either end of the path. Another likely suspect is interference.

Share and Share Alike

The AX.25 packet system uses a frequency-sharing scheme based on the ability of your TNC to sense other transmissions. Random time delays are used as well. The result is that several stations can occupy the same frequency, each taking its turn to transmit packets.

If one station is transmitting, all the other TNCs "hold their tongues" until the transmission is finished. When the frequency is silent, the TNCs pause before they attempt to transmit. The length of each TNC's pause is determined by a

Packet Tips: Stream Switching

Did you know you can talk to several packet stations at once? Believe it or not, most TNCs are capable of handling 10 or more separate conversations through a technique known as *stream switching.*

Enter the command mode and take a peek at your **USER** parameter. If it's set at 1, your TNC will accommodate only 1 connection at a time. Anyone who attempts to connect to you when you're already connected to someone else will receive a *busy* message followed by an immediate disconnect. By setting the **USER** parameter to a number greater than 1, you're increasing the number of stations that can connect to you simultaneously. The setting of your **MAXUSERS** parameter also determines how many streams will be available.

Your TNC will treat each separate connection as a different channel, or *stream.* Each stream is labeled with an alphabetical letter (stream A, B, C and so on). To switch from one stream to another, you must enter the stream-switch character (l in many TNCs) followed by the alphabetical letter for the stream you want to use. You *do not* need to be in the command mode to switch streams.

Juggling multiple conversations is quite a challenge, but it's fun, too. You haven't lived until you've experienced the frenzy of jumping from stream to stream, answering questions or sending comments on the fly! There's a practical side to stream-switching, too. What if another ham needs to talk to you *right now*? If you've set up your TNC for multiple connections, the impatient amateur can still connect—even while you're talking to someone else. Your TNC will let you know that another connection has taken place and you can quickly switch to the appropriate stream without missing a beat! Various TNC models handle stream switching in different ways. Take a few minutes to read your manual before leaping into the fast lane of multi-stream conversations.

set of random-number calculations. It's like the game of breaking twigs into several pieces to see who draws the shortest stick. In the case of packet, the TNC that calculates the shortest delay transmits first. All the other TNCs must wait until the frequency is silent once again. At the next opportunity, more delays are randomly calculated and, with luck, a different station will be the "winner."

If the system works properly, the possibility of two stations transmitting at once is minimal. Everyone gets a fair chance to transmit and the data flows from station to station with reasonable efficiency—although the transfer rate, or *throughput*, decreases as the frequency becomes crowded. The major weakness of this system is it only works when *all* stations can hear each other. As you can probably guess, this isn't always the case!

Suppose another station arrives on the frequency. Despite its strong signal, it can't hear you or the station you're connected to. As far as its TNC is concerned, the frequency is completely clear. Since it hears no other signals, it transmits whenever it chooses, wreaking havoc on your packet connection. It's similar to the frustration of trying to carry on a conversation at an airport. Every time an airliner roars by, you and your friend are reduced to shouting, "What? What did you say?"

When Your TNC Gives Up

If interference is severe, or propagation is poor, how long will your TNC keep trying before it finally gives up? It all depends on the **RETRY** parameter. Your TNC was probably shipped with its **RETRY** parameter set at 10. In other words, it will transmit a packet 10 times without acknowledgment before terminating the connection. You *can* set your **RETRY** parameter to zero, forcing your TNC to repeat its packets endlessly, but this is poor operating practice. If the signal path between you and another station is so poor that it takes 20 or 30 "retries" to get a packet through, why bother? Unless it's an emergency situation, you'll only cause frustration for yourself and additional congestion for everyone else.

In the end it doesn't matter if it's interference or poor propagation that eventually pushes your TNC to its limit. When the **RETRY** count is exceeded, you'll see:

Retry count exceeded
***DISCONNECTED

This is a frustrating sight to any packeteer, but what's the solution? You could erect huge directional antennas and run enormous amounts of power, or . . . you could find another station to relay packets for you. With a relay in the right location, even a low-power station with marginal antennas could make connections over a large area. It sounds like I'm describing an FM voice repeater, doesn't it? If you answered "yes," you're close! In packet radio, however, we call these special devices *nodes* and *digipeaters*.

Packet Nodes and Digipeaters

It's easy to confuse nodes and digipeaters with voice repeaters. There are similarities, but there are major differences as well . . .

❏ Voice repeaters listen on one frequency and transmit on another. Most nodes and digipeaters listen and transmit on the *same* frequency. The exceptions are nodes that act as links to other stations via UHF or VHF *backbone* systems, or nodes that function as *gateways* to and from the HF bands.

❏ Voice repeaters relay *every* signal heard on the input frequency. Nodes and digipeaters relay only packets addressed specifically to them.

❏ Voice repeaters are *real-time* devices. They listen and transmit simultaneously. Nodes and digipeaters receive packets, store them momentarily and then retransmit.

In the early days of amateur packet, nodes didn't exist. If you needed a relay, you used a digipeater. Every second-generation TNC (called a *TNC2*) had the ability to function as a digipeater, so any station could act as a relay for another. (This is still true today.) Some amateurs established powerful digipeaters by placing TNCs, transceivers and antennas at excellent operating sites. This allowed hams in poor locations to extend their coverage.

To use another station as a digipeater, you have to let your TNC know that the connecting path is not direct. In other words, you need to change your usual connect request to specify which station is acting as the digipeater relay.

CONNECT W7FUR VIA WA7MSO

In this example, your TNC will try to connect to W7FUR using WA7MSO as a digipeater. If more than one digipeater is needed to make the connection, you must list them in proper order. The digipeater nearest to you appears immediately after the "via." The digipeater nearest the destination station is shown last.

CONNECT W7FUR VIA WA7MSO, W7GHE, KA7ZZP

The big problem with digipeaters is that they aren't "intelligent." They shuffle data from one station to another and nothing else. If you connect to a station through a digipeater, every bit of information—including every ACK—has to travel from one end of the path to the other (see Fig 4-1). The longer the path, the greater the possibility of failure.

NET/ROM nodes brought major changes to packet radio. Unlike digipeaters, nodes are sophisticated data-handling devices. A particularly complex node is able to send and receive packets on several different frequencies. Each input/output frequency is known as a *port* and it's not unusual for a single node to have ports on the HF, VHF, UHF and microwave bands! This enables the node to perform its relays

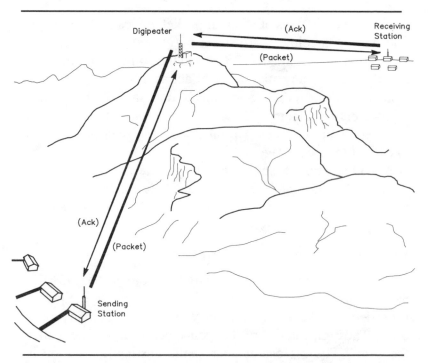

Fig 4-1—Digipeaters can be used to span the gaps between distant stations. However, every bit of packet data must travel back and forth through the digipeater from one station to the other. In this example, the packet and the ACK must complete the entire journey from end to end.

on whatever frequencies are required to get packets to their destinations.

When you send packets to a node, the node assumes complete responsibility for relaying them to the target station, or to another node. There's no need to list nodes in your connect requests. Nodes also eliminate the inefficiency of end-to-end acknowledgments. If the destination station receives your packets error-free, its ACK only has to travel to the nearest node—not all the way back to your station (see Fig 4-2)!

No matter where you live, there's probably a *NET/ROM*-

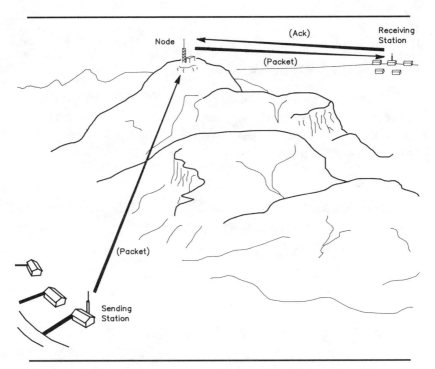

Fig 4-2—Packet nodes are more efficient than digipeaters. The node relays the data in this example, but the ACK from the receiving station only needs to travel back to the node.

compatible node within range of your station. There are other types of nodes as well, which we'll discuss in Chapter 6. If you know the call signs of other hams on the frequency, you can still use their stations as digipeaters when necessary (assuming that they haven't turned their digipeater functions off). I think you'll find, however, that nodes are more efficient devices for relaying packets.

Let's pretend that you want to contact KC4WF, but the signal path between your stations won't support a direct connection. Here's where a node proves its true worth! You begin by establishing a connection to a nearby *NET/ROM* node. We'll call it W4ILW-5.

CONNECT W4ILW-5

If the node acknowledges your request, you'll see:

*** CONNECTED to W4ILW-5

Now you can use the node to reach KC4WF. When the connection was established, your TNC switched from the command mode to the converse mode. *Without leaving the converse mode*, simply enter:

C KC4WF

You've asked the node to contact KC4WF for you. With any luck you'll be rewarded with:

*** CONNECTED to KC4WF

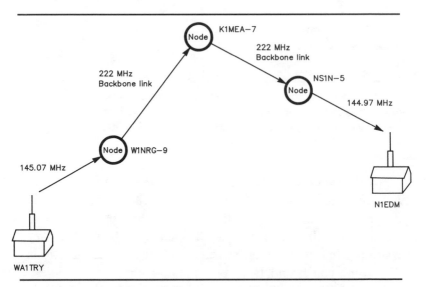

Fig 4-3—It's possible to use several nodes to form a path from one station to another. In this example, WA1TRY connects to the W1NRG-9 node on the 2-meter band. W1NRG-9 relays the data via a 222-MHz backbone link to K1MEA-7 and NS1N-5. The destination station, N1EDM, receives the data on 2 meters from the NS1N-5 node.

You can chat with KC4WF as long as you like and the W4ILW-5 node will handle all the relays. If a packet needs to be transmitted more than once between the node and the target station, the node will do so automatically.

If the distance between you and another station is beyond the capability of one node to handle alone, you can link several nodes together to form a multi-node path. All you need to know is the call sign of the node nearest the station you want to contact. (Let's call that node K4MEA-7.) Connect to your local node and send the **NODES** command, or simply **N**. You've just asked for a list of all the nodes your local node has heard recently. As the node responds to your request, you'll see something like this:

RIVER:W4ILW-5} Nodes:

SAVAN:WA4TPP-3	SAVAN2:WA4TPP-13	CENTGA:K4BKE
GRNVL:WA4TPP-2	LOUIS:K4EIC-2	SOUFL:WA4UQC-7
MONT:N4API-5	SPFLD2:W4NY-2	WESTNC:K4MEA-7

The call sign of each node in the list is preceded by its alias. Do you see K4MEA-7 in the list? Good. This means that your local node may be able to establish a link with K4MEA-7. The easy way to find out is to try it!

You're already connected to W4ILW-5. *Remain in the converse mode* and issue a connect request for K4MEA-7.

Packet Tips: Aliases

In addition to their call signs, packet nodes can also be identified by their *aliases*. Aliases come in handy when you can't seem to remember the call sign of the node you want to reach. They often indicate the location of the node or the name of the group that operates it. For example, the alias of W4ILW might be RIVER—an abbreviation for Riverside. You can command your TNC to connect to W4ILW-7 *or* RIVER. The connection will be made regardless of whether you use the call sign or the alias!

(This request is being sent to the node, not as a command for your TNC.)

C K4MEA-7

Sit back and relax. The nodes will take care of everything—even using 222- or 420-MHz links if necessary to reach the destination. If the nodes are able to establish a path to K4MEA-7, you'll see a message similar to the one shown below.

W4ILW-5> Connected to K4MEA-7

All you have to do is issue your final connect request to the station you wish to contact:

CONNECT N4ATQ

It's important to point out that QSOs over long paths (linking more than two or three nodes) are discouraged in

---

Packet Tips: Using the ROUTES Command

When you ask a *NET/ROM* node for a list of all the nodes it can reach, does this mean you can establish a connection to every node on the list? Not necessarily!

Node lists are updated automatically. If transmissions from a new node are received, it may earn a place on the list. A well-managed system will only list a node if its signal quality is sufficient to guarantee a reliable connection. Consistency is important, too. If a node makes the list but isn't heard again within a certain period of time, it's removed.

Node operators are hams just like you. Some are very attentive to their systems. Others are busy with work or family matters and have little time to maintain their nodes. If you send the **NODES** command and receive a long list of various nodes, be suspicious! A poorly managed node will include virtually everything it hears, regardless of signal quality.

---

some areas because they tend to reduce the efficiency of the network. Longer paths are also unreliable and prone to failure.

Packet Bulletin Boards

Packet bulletin board systems (PBBSs) form the hubs of most VHF packet networks and they're often found on the HF bands as well. PBBSs are electronic warehouses for the bulletins, private mail and NTS traffic that flow through the packet system. By connecting to a PBBS, you'll be able to read a variety of bulletins, send mail to other packet-active hams or send traffic to just about anyone!

Most PBBSs are operated by clubs or private individuals. It takes a fair amount of time and money to maintain such a system. The system operator—or SysOp—is the person who calls the shots on any PBBS. After all, it's his time and, in many cases, his money! The worldwide packet network

So how can you know if the path to a node is reliable? Try using the **ROUTES** command. Depending on the software the node is using, you'll see a response similar to this:

MARC:W1NRG-9} Routes:

Port	Neighbor Node Call	Quality
1	#MMK13:N1API-13	155
1	MARS:W1EDH-3	100
1	SPFLD2:W1NY-2	100
1	BERK2:WA1TPP-13	155
1	#NWCT7:WA1UQC-8	155

Check the numbers in the **Quality** column. Generally speaking, a quality figure of 100 or more is reliable (255 is the highest quality possible). If the node you're interested in reaching shows a quality *below* 100, your chances of success are probably poor.

This is the W1NRG PBBS operated by the Meriden (Connecticut) Amateur Radio Club. Directly beneath the monitor you'll notice two packet TNCs and two transceivers. One TNC/transceiver combination is for the 2-meter user input on 144.97 MHz. The other TNC/transceiver combo is for the 222-MHz backbone link. An RF amplifier and power supply are also visible.

depends upon the dedication of SysOps. When you consider that this is an all-volunteer effort, it is amazing that the system functions as well as it does.

Connecting to a Bulletin Board

Connecting to a PBBS is the same as connecting to any other packet station. Monitor the packet frequencies and watch for bulletin board activity. When you find one, make your connect request and have fun! Let's pretend we've discovered the W1NRG-4 PBBS.

CONNECT W1NRG-4

We'll also presume that you're close enough to connect directly. If not, you'd use a node to bridge the gap.

*** CONNECTED TO W1NRG-4

Packet Tips: Help Files

When you connect to a PBBS for the first time, check its file directory and see if a *help file* is available. Having a help file at your fingertips will save you a great deal of time and frustration when you need to use a particular PBBS function.

Begin your search by sending the letter **W**.

The PBBS will respond with a list of all files available for downloading to your computer. Look for a file name that includes the word "help." If you don't see a file with this label, look for "INFO" or "PBBSINFO." Transferring the file to your computer is easy. Just send the following:

D <file name>

D stands for download and "file name" is the complete name of the file. If your terminal software has a *capture* or *ASCII save* function, turn it on quickly so that all the incoming information will be saved to a disk file.

Some sophisticated PBBSs store their files in separate directories. When you send the **W** command, you'll be presented with a list. To see what a directory contains, you'll need to send:

W <directory name>

To download a file from a directory, you need only modify the command slightly. . .

D <directory name>\<file name>

Once you've downloaded the help file, print it out and keep it handy for future reference. If you can't locate the help file, ask the SysOp for assistance.

Very good! You've made the connection. Now the PBBS will send its sign-on information to you.

**Welcome to W1NRG's MSYS PBBS in Wallingford, CT
Enter command:
A,B,C,D,G,H,I,J,K,L,M,N,P,R,S,T,U,V,W,X,?,* >**

Table 4-2

Common PBBS Commands

General Commands:

B	Log off PBBS.
Jx	Display call signs of stations recently heard or connected on TNC port x.
N x	Enter your name (x) in system (12 characters maximum).
NE	Toggle between short and extended command menu.
NH x	Enter the call sign (x) of the PBBS where you normally send and receive mail.
NQ x	Enter your location (x).
NZ n	Enter your ZIP Code (n).
P x	Display information concerning station whose call sign is x.
S	Display PBBS status.
T	Ring bell at the Sysop's station for one minute.

Information commands:

? *	Display description of all PBBS commands.
?	Display summary of all PBBS commands.
? x	Display summary of command x.
H*	Display description of all PBBS commands.
H	Display summary of all PBBS commands.
H x	Display description of command x.
I	Display information about PBBS.
I x	Display information about sation whose call sign is x.
IL	Display list of local users of the PBBS.
IZn	List users at ZIP Code n.
V	Display PBBS software version.

Message commands:

K n	Kill message numbered n.
KM	Kill all messages addressed to you that you have read.
KT n	Kill NTS traffic numbered n.
L	List all messages entered since you last logged on.
L n	List message numbered n and messages numbered higher than n.
L< x	List messages *from* station whose call sign is x.
L> x	List messages addressed *to* station whose call sign is x.
L@ x	List messages addressed for forwarding to PBBS whose call sign is x.

L n1 n2	List messages numbered n1 through n2.
LA n	List the first n messages stored on PBBS.
LB	List all bulletin messages.
LF	List all messages that have been forwarded.
LL n	List the last n messages stored on PBBS.
LM	List all messsages addressed to you.
LT	List all NTS traffic.
R n	Read message numbered n.
RH n	Read message numbered n with full message header displayed.
RM	Read all messages addressed to you that you have not read.
S x @ y	Send a message to station whose call sign is x at PBBS whose call sign is y.
S x	Send message to station whose call sign is x at this PBBS.
SB x	Send a bulletin message to x at this PBBS.
SB x @ y	Send a bulletin message to x at PBBS whose call sign is y.
SP x @ y	Send a private message to station whose call sign is x at PBBS whose call sign is y.
SP x	Send a private message to station whose call sign is x at this PBBS.
SR	Send a message in response to a message you have just read.
ST x @ y	Send an NTS message to station whose call sign is x at PBBS whose call sign is y.
ST x	Send an NTS message to station whose call sign is x at this PBBS.

File transfer commands:

Dx y	From directory named x, download file named y.
U x	Upload file named x.
W	List what directories are available.
Wx	List what files are available in directory named x.
Wx y	List files in directory named x whose file name matches y.

That's a big list of commands and we don't have the space in this book to discuss each one. A list of common PBBS commands is shown in Table 4-2.

Let's try some of the popular functions instead. Would you like to see some of the bulletins available on the PBBS? The *LB* (List Bulletins) command will do the trick!

MSG #	TR	SIZE	TO	FROM	@BBS	DATE	TITLE
744	B#	473	WANTED	K1CC	USBBS	920920	AT motherboard
743	B#	654	ALL	N1GFL	NEBBS	920920	HF RIG FOR SALE
739	B#	652	NEEDED	KO1C	USBBS	920919	NEED DENTRON MANUAL
736	B#	488	WANT	KA1TYV	USBBS	920919	SOFTWARE HP 100/150
733	B#	853	INFO	N1IPO	USBBS	920919	WHERES MY CALL SIGN?
730	B#	949	HELP	WA3YHH	USBBS	920919	MODS TO IC-22A
710	B#	5872	SWL	AE1T	NEBBS	920919	English Broadcasts
708	B#	611	ALL	W1OER	NEBBS	920919	FOR SALE:ICOM 228H

You can read any of these messages by sending the letter *R* followed by the message number. Sending "R 743" allows you to read message number 743 concerning an HF rig for sale.

Sending Packet Mail

Would you like to send a message to a packet-active amateur in another city, state or country? If you know the call sign of his local PBBS, it's easy!

SP N6ATQ @ WA6ZWJ

This odd-looking line translates to: Send a personal message (SP) to N6ATQ at (@) the WA6ZWJ bulletin board. The PBBS now asks for the subject of the message:

SUBJECT?:

Enter a subject sentence. Keep it *very* short, preferably less than 28 characters (including spaces).

WORKED MEXICO ON 6 METERS!

Packet Tips: Hierarchical Addressing

Hierarchical addressing sounds like a complex concept, but you probably do it every day! When you address an envelope, you begin with the name of the addressee followed by the street and house (or apartment) number. After that you add the city and finally the state. Whether you know it or not, you've created a multi-level *hierarchical* address that begins with a very specific location (the house number) and ends with a very general location (the state).

Packet mail addresses can also be hierarchical. For example:

WB8IMY @ W1NRG.CT.USA.NA

This hierarchical address begins with my call sign and then goes on to list my local PBBS (W1NRG), my state (Connecticut), my country (USA) and finally the continent where I live (North America, or NA).

Hierarchical addressing helps packet mail travel more efficiently through the global network. By using hierarchical addressing for all your personal mail, the odds for successful delivery will be improved.

The PBBS will respond with something like this:

ENTER MESSAGE. USE CTRL-Z OR /EX TO END MESSAGE

That's your cue to begin entering the text of your message. If the message is going to travel out of your region, keep it as brief as possible. Longer messages travel much slower. When you've entered your message, enter CONTROL-Z or /EX on a line by itself. You'll soon see the PBBS *command menu* again.

Enter command:
A,B,C,D,G,H,I,J,K,L,M,N,P,R,S,T,U,V,W,X,?,* >

How long will it take for the message to reach your friend? It depends on a host of factors, including propagation.

The packet network is not a commercial system. You can't expect the same level of dependability. Your message may arrive in a few hours, a few days, a few weeks or not at all! Even so, packet mail is surprisingly reliable and efficient. I can usually send messages from Connecticut to Ohio, for example, in about 12 hours or less.

Sending a Bulletin

Sending a bulletin is similar to sending private mail. The only difference is that you don't have a particular person or destination in mind. Let's say you need help with a problem and you live in the state of Montana.

SB ALL @ MTBBS

This means: Send a bulletin (SB) to everyone (ALL) at (@) every PBBS in Montana (MTBBS).

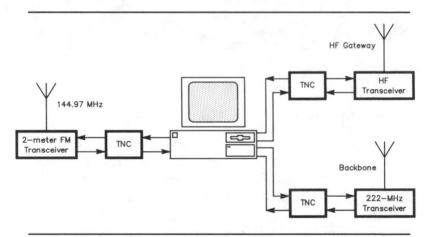

Fig 4-4—This is an example of a typical *multi-port* PBBS. Local users connect to the system on the 2-meter band. Packet mail is sent on the 222-MHz backbone link (for local and regional destinations), or via the HF gateway (for distant stations). This system is said to have three ports: 2 meters, 222 MHz and HF.

Packet Tips: The Mail Must Get Through

All personal messages, bulletins and NTS traffic are sent through the global packet network using a process known as *forwarding.* Messages are passed from one PBBS to another as they make their way through the system. Forwarding can take place on a variety of bands depending on the destination of the message or the available frequencies in the network.

In less active areas, PBBSs forward mail on the same bands they use for normal user access. In areas where packet is very active, however, you'll find PBBSs forwarding on the 222- and 420-MHz bands. These are known as *backbone* links. Their only purpose is to carry mail from one PBBS to another—often at data rates of 9600 bits per second or higher. In extremely crowded areas, hams are operating high-speed backbones on the microwave bands.

If packet mail must be sent over long distances, forwarding is usually conducted on HF frequencies at some point in the process. If you monitor HF packet, you'll see mail forwarding taking place.

In the "old days," mail forwarding was a major operation for a PBBS. The system would shift into the forwarding mode and all normal functions would be suspended until the forwarding was complete. (This could take as long as an hour if a PBBS was sending or receiving a lot of mail!) For the users it meant receiving "busy" messages when they tried to connect.

As packet has evolved, *multi-user* PBBSs have become common. Not only can a multi-user system handle several connections at once, it can conduct forwarding operations at the same time! Recently I posted a message to a friend in another city, and then spent a few minutes listing and reading bulletins. When I checked to see if my message was still there, I was surprised to find it had vanished. The multi-user PBBS had forwarded my message while I was still connected to the system!

Packet Tips: Who Should See Your Bulletins?

Sending packet bulletins is a great way to get the attention of many hams at once. If you have a problem, an announcement, an item for sale or an item you'd like to buy, by all means put it on packet! Before you enter that **SB** command, however, consider the scope of your message.

It's always tempting to go for maximum distribution (to "ALL @ USBBS"), but is that really what you want? Is it in your best interests, and the best interests of the packet community, to have your message appear at every PBBS in the United States? Here are some guidelines:

❑ If you're sending a WANTED or FOR SALE bulletin, send it to the PBBSs in your home state *first*. If you live in Ohio, you'd enter **SB ALL @ OHBBS**. Wait a week to see what happens. If you don't get a response, send it to the PBBSs in your *region*. To send the bulletin to every system in the northeast, for example, you'd enter **SB ALL @ NEBBS**. Only when this fails should you consider the USBBS option.

❑ Will amateurs living outside your state or region be interested in the content of your message? If you're announcing a schedule for license examinations at a hamfest in New Jersey, it doesn't make sense for that message to appear on PBBSs in California. A 200-ft tower for sale in Maine will probably be of little interest to a ham in Texas.

❑ If you're using the USBBS option to announce an event that will take place on a particular date and time, send it early! Remember that it may take as long as two weeks for a general bulletin to reach every PBBS in the country. It's always frustrating to read about an interesting event several days after the fact!

SUBJECT?

When you see this prompt, describe your problem in as few words as possible.

HELP! ANTENNA ROTATOR STUCK!

As before, enter your text and end your message. Connect to the PBBS a few days later and you may have some helpful mail waiting! You can also use a packet bulletin to announce that you have an item for sale. When this book went to press, FCC rules specified that items for sale must be directly related to Amateur Radio activity (a transceiver, an antenna, a computer, a monitor and so on). These regulations may be revised, however. If you're in doubt, contact the ARRL Regulatory Information branch.

Finally, if you have something to say about a specific topic (shortwave listening, for example), you can address your bulletin accordingly:

SB SWL @ MTBBS

This format is particularly helpful to packeteers who use programs that scan for keywords in a bulletin list and automatically download items of interest.

Sending NTS Traffic

NTS traffic can be originated and received at most packet bulletin boards. Not only is this useful in emergencies, it's also handy for communicating with nonhams. Let's say I want to send an NTS message to my parents. I'd begin by entering:

ST 45429 @ NTSOH

This means: Send traffic (ST) to my parents' ZIP code (45429) at (@) National Traffic System/Ohio (NTSOH). Standard two-letter state postal abbreviations are used. If my parents were in California, I would send it to NTSCA.

When the PBBS asks for a subject, I respond with:

QTC Kettering (513) 555

This cryptic line simply translates to: Traffic for (QTC) the city of Kettering with the telephone area code and prefix as shown. Now I enter my message in NTS packet format. The first line contains information about the nature of the message:

NR 3 ROUTINE WB8IMY 23 WALLINGFORD CT DEC 18

Can you guess what this means? It's not as complicated as it looks: Message number 3, routine priority, from WB8IMY, with a total of 23 words, originating from Wallingford, Connecticut, on December 18. Now I have to supply the name, address and complete telephone number of the recipient.

Mr and Mrs Robert Ford
549 Shady Lane
Kettering OH 45429
513 555 3958

Finally, I enter my message:

Sending you a special Christmas surprise   Should arrive via UPS on Wednesday or Thursday   Hope you enjoy it Will call Christmas day

Love   Steve and Kathy

That's all there is to it! The packet bulletin board will relay the message to the network as soon as possible. Did you notice the total lack of punctuation in the message? This is because the message may be intercepted and relayed on CW or phone nets where punctuation is not often used. Either way, an amateur in the Kettering area will call my parents and deliver the happy message.

You can deliver traffic, too! Check your PBBS for traffic using the **LT** (*List Traffic*) command. As you read the list of

Packet Tips: Should You Start Your Own PBBS?

Are you thinking about starting your own packet bulletin board system? Take this simple test first.

❏ Do you own a spare computer with a sizable hard disk that you can dedicate completely to the PBBS?

❏ Do you own, or are your prepared to buy, the necessary PBBS software?

❏ Will you maintain your system in good operating condition and make it available to users 24 hours a day, 365 days a year?

❏ Do you own a TNC that you can dedicate to the operation of the PBBS?

❏ Are you prepared to dedicate a transceiver and antenna system to the PBBS? Will your station have sufficient output power and receive sensitivity to allow reliable access over a wide area?

❏ If mail forwarding is conducted on other bands, do you have, or will you be able to purchase, the necessary equipment to access the backbone links (a separate transceiver, antenna and TNC)?

❏ And finally, are you certain that your area *needs* another PBBS? (Will your system duplicate the work of another? Will it interfere with other systems on the same frequency?)

If you answered "yes" to all of these questions, you're probably ready to become a PBBS SysOp. Check the Packet Info Guide for information on PBBS software and packet equipment.

messages, do you see any intended for your area? If so, use the **R** command to read the traffic. Write the message down or print it on your printer. If you decide to deliver it, use the **KT** (*Kill Traffic*) command to delete the message from the

PBBS. Delivering NTS traffic is a pleasurable experience and may turn you into a dedicated traffic handler!

Other PBBS Functions

Depending on the sophistication of the system, a PBBS may offer other functions you'll find useful. Call sign directories are available on CD-ROM and many PBBSs are beginning to incorporate these into their systems. I can look up domestic call signs on my local PBBS by sending the **PC** command. For example, if I send **PC WB8ISZ** the PBBS will respond with the complete address of WB8ISZ!

You can download text files from a PBBS as well. The **W** command will show you what the PBBS has to offer. Transferring a file to your computer is as easy as using the **D** (download) command. (See the sidebar "Packet Tips: Help Files.") Most systems offer only ASCII text files for downloading. Binary files—computer programs, for example—are sometimes available, but require specialized software to transfer them to your computer.

DX PacketClusters

Not all areas of the country are blessed with *Packet-Clusters*, but their population is growing rapidly. You may stumble upon one as you're monitoring your local packet frequencies. At first glance it may look like a bulletin board, but it isn't!

A *PacketCluster* is a network of nodes operating under specialized *PacketCluster* software. They're dedicated to contest and DX activities. You connect to a *PacketCluster* in the same way you'd connect to any other station. However, the information you'll receive will be very different!

*** CONNECTED to KC8PE

Welcome to YCCC PacketCluster node - Cheshire CT
Cluster: 22 nodes, 9 local / 144 total users Max users 367

WB8IMY de KC8PE   5-May-1992 0124Z   Type H or ? for help>

The cluster is waiting for your command. What would

you like to see? How about a list of the latest DX sightings (called *spots*)? You only need to enter: **SHOW/DX**

7015.5	**UL7MG**	**5-May-1992**	**0119Z**	**<W3XU>**
14029.3	**UD6DFF**	**5-May-1992**	**0106Z**	**<K2LE>**
14211.9	**J68AJ**	**5-May-1992**	**0056Z**	**<NE3F>**
14002.6	**4S7WP**	**5-May-1992**	**0054Z**	**<K2LE>**
14007.0	**ZA1ED**	**5-May-1992**	**0048Z**	**<K2LE>**

WB8IMY de KC8PE   5-May 0124Z >

Now you have a list of the five most recent DX spots along with their frequencies and the times (in UTC) when they were heard. If you're tuning through the 15-meter band, for example, you may discover another DX station worthy of a spot on the cluster. Go ahead and make a contribution by posting it on the network. The simplest command format would be: **DX 21.250 SV3AQR**

PacketClusters have other useful features. You can use the **DIR** command to see a list of bulletins just as you would on a packet PBBS. Using the **R** (READ) command will allow you to read any bulletin you wish. Unlike packet bulletin boards, however, *PacketClusters* can relay bulletins and messages only within the networks they serve.

Sending the **SHOW/WWV** command will give you the latest solar activity reports.

Date	Hour	SFI	A	K	Forecast
27-Apr-1992	21	137	4	2	Low/Quiet =>Low-Mod/Qt-Uns
27-Apr-1992	12	143	6	0	Low/Quiet => Low-Mod/Qt-Uns
27-Apr-1992	09	143	6	1	Low/Quiet => Low-Mod/Qt-Uns
27-Apr-1992	06	143	6	0	Low/Quiet => Mod/Uns
27-Apr-1992	03	143	6	2	Low/Quiet=>MstlyMod/Settled

In this case I can see that the A and K index numbers are low, which is good, but solar flux (SFI) numbers in the 137 to 143 range indicate only fair DX conditions. On the other hand, low A and K numbers combined with a high solar flux (200 or higher) would promise very good DX hunting on the bands! As I'm watching the *PacketCluster*, I see a spot for a

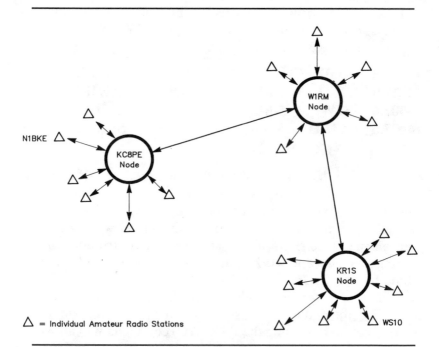

Fig 4-5—*DX PacketClusters* are networks comprised of individual nodes and stations. In this example, N1BKE is connected to the KC8PE node. If he finds a DX station on the air, he'll post a notice—otherwise known as a *spot*—which the KC8PE node distributes to all its local stations. In addition, KC8PE passes the information along to the W1RM node. W1RM distributes the information and then passes it to the KR1S node, which does the same. Eventually, WS1O—who is connected to the KR1S node—sees the spot on his screen. Depending on the size of the network, WS1O will receive the information within minutes after it was posted by N1BKE.

DXpedition on Sable Island. Hmmm . . . I could use that one! I wonder what bands would provide the best propagation from my location to Sable Island? Why not ask the *PacketCluster*? All I have to do is send: **SHOW/M CYØ**

M stands for *maximum usable frequency* (or *MUF*) and

CYØ is the call sign prefix for Sable Island. Here's how the *PacketCluster* responds:

Sable-Is Propagation: Flux: 137 Sunspots: 90
Rad Angle: 29   Dist: 995 km   Hops: 1
MUF (90%): 9.0 (50%): 10.7 (10%): 13.1

The MUF calculations tell me that propagation to the island is ideal at 9 MHz (90%), but drops rapidly as the frequency increases. The 30-meter band would be best, but they're only working stations on 20 meters. Sure enough, the prediction of marginal propagation on 20 meters was correct. (I still managed to work them, though!)

DX PacketClusters also offer you the capability of chatting with other stations connected to the cluster. You can send a simple greeting by using the **TALK** command:

TALK NF1J   Hello, Warren. Nice job working 5U7M!

NF1J will see my greeting on his screen within a minute or two after I send this line, depending on network activity. If he wants to talk to me at length, he can use the **TALK** command to establish a link between our stations. The *PacketCluster* will continue to show us new DX sightings as they appear, but everything we type will be sent to each other.

If you want to see who's connected to the network, just send the **SHOW/C** (show configuration) command. The *PacketCluster* will respond with a complete list of every station connected to the network grouped by the node they're using. Here's a typical example:

PacketCluster Configuration:

Node	Connected stations			
KC8PE	N1API	KS1L	K1WJL	KC1SJ
	WB8IMY			
W1RM	(WB1AIU)	NTØZ	KA1BSA	AB1U
	KB1LE	N1JBH	N1GLA	KG1D-1

		(NJ2L)	(NA1I)	KB1HY	K1FRD
		NX1L	KB1BE	W1CKA	KB1CQ
		(W1GG)	K1ZJH	WB1GUY	K1KI
K2TR	KA2EXB	(K2QE)	N2JJ	NJ1F	
	K2VV	(KB2HUN)	K2ONP	KQ2K	
	WK2H	KA2HTU	WS2U	N2EKU	

You can send TALK messages, or enter into a conversation with any ham on the list. The only exceptions are the call signs in parenthesis. These hams are connected to the cluster, but away from their keyboards temporarily. How does the system know this? Actually, it doesn't—unless you tell it. Yes, there is a command for that, too! And like PBBSs, you don't need to send the full command every time. Instead, you can use the abbreviated form such as SH/DX rather than SHOW/DX. Take a look at the *PacketCluster* commands shown in Table 4-3.

Some *PacketCluster* networks include QSL manager data bases. As soon as you've worked your latest DX station, it's a good idea to check and see if a QSL manager is available. I just worked the Sable Island DXpedition, so I'll use the data base to find out where to send my card. All I have to do is enter: **SHOW/QSL CYØSAB**.

WB8IMY de KC8PE   28-Apr 0019Z >
Accessing remote data base on KB1H . . . standby . . .

WB8IMY de KC8PE   28-Apr 0019Z >
The W6GO/K6HHD QSL Manager list #145, 20-Apr-92, licensed for use on the KB1H node, says the address as printed in QTH-Iss.141 for CYØSAB is:

VE1CBK, PO Box 32, Site 35, RR#1, Windsor Jct, N.S. CANADA B0N 2V0

Type SH/QSL SUBS for a list of those who provide this data for you. Additional information to add? Type

UPDATE/QSLNEW/APPEND.
WB8IMY de KC8PE   28-Apr 0020Z >

My card will be on its way to Canada in tomorrow's mail!

Specialty Nodes

Packet radio is constantly evolving and improving. As it does so, new features appear. Perhaps you'll find a *conference* node or *chat* node. These systems function as gathering spots for amateurs who want to shoot the breeze with each other. As we saw earlier, you can carry on multiple conversations by using stream switching, but conferencing takes this idea one step further and makes it much easier for everyone. All you have to do is connect to the conference node and you're instantly plunged into a packet conference. Whenever someone sends a message, it's seen by all the other stations. Joining a conference is like attending a party on the air!

As you explore packet radio, you may also stumble upon a *weather node*. It sounds incredible, but a weather node is a packet station that provides real-time weather telemetry to anyone who connects.

Welcome to the N1HUI Weather Node, Branford, Ct
H/Help D/Data

If I send a **D** to see data, this weather node responds with:

01/26/92 20:23:14 TF	=	+0027.9 DEGF
01/26/92 20:23:14 TPCB	=	+0080.5 DEGF
01/26/92 20:23:14 WD	=	00340 DEG NNW
01/26/92 20:23:14 WS	=	00018 MPH

I can see that the outside temperature (TF) is a brisk 27.9 degrees Fahrenheit. The indoor temperature (TPCB), is a cozy 80.5 degrees Fahrenheit. The wind is coming out of the north-northwest (WD) at 18 miles per hour (WS).

Table 4-3

Common *DX PacketCluster* Commands

ANNOUNCE	Make an announcement.
A *x*	Send message x to all stations connected to the local node.
A/F *x*	Send message x to all stations connected to the cluster.
A/*x y*	Send message y to stations connected to node x.
A/*x y*	Send message y to stations on distribution list x.
BYE	Disconnect from cluster.
B	Disconnect from cluster.
CONFERENCE	Enter the conference mode on the local node.
CONFER	Enter the conference mode on the local node. Send <CTRL-Z> or /EXIT to terminate.
CONFER/ F	Enter the conference mode on the cluster. Send <CTRL-Z> or /EXIT to terminate.
DELETE	Delete a message.
DE	Delete last message you read.
DE *n*	Delete message numbered n.
DIRECTORY	List active messages on local node.
DIR/ALL	List all active messages on local node.
DIR/BULLETIN	List active messages addressed to "all."
DIR/*n*	List the n most recent active messages.
DIR/NEW	List active messages added since you last invoked the DIR command.
DIR/OWN	List active messages addressed from or toyou.
DX	Announce DX station.
DX *x y z*	Announce DX station whose call sign is x on frequency y followed by comment z, e.g., DX SP1N 14.205 up 2.
DX/*a x y z*	Announce DX station whose call sign is x on frequency y followed by comment z with credit given to station whose call sign is a, e.g., DX/K1CC SP1N 14.205 up 2.
FINDFILE	Find file.
FI *x*	Ask the node to find file named x.
HELP or ?	Display a summary of all commands.
HELP x	Display help for command x.
READ	Read message.

R	Read oldest message not read by you.
Rn	Read message numbered n.
R/*x y*	Read file named y stored in file area named x.
REPLY	Reply to the last message read by you.
REP	Reply to the last message read by you.
REP/D	Reply to an delete the last message read by you.
SEND	Send a message.
S/P	Send a private message.
S/NOP	Send a public message.
SET	Set user-specific parameters.
SE/A	Indicate that your computer/terminal is ANSI-compatible.
SE/A/ALT	Indicate that your computer/terminal is reverse video ANSI-compatible.
SE/H	Indicate that you are in your radio shack.
SE/L *a b c d e f*	Set your station's latitude as: a degree b minutes c north or south and longitude d degrees e minutes f east or west, e.g., SE/L 41 33 N 73 0 W.
SE/N x	Set your name as x.
SE/NEED x	Store in database that you need country(s) whose prefix(s) is x on CW and SSB. e.g., SE/NEED XX9.
SE/NEED/BAND =(*x*)y	Store in database that on frequency band(s) x, you need country(s) whose prefix(s) is y, e.g., SE/NEED/BAND=(10)YA.
SE/NEED/*x y*	Store in database that in mode x (where x equals CW, SSB or RTTY), you need country(s) whose prefix(s) is y, e.g., SENEED/RTTY YA.
SE/NEED/x/BAND =(y)z	Store in database that in mode x (where x equals CW, SSB or RTTY) on frequency band(s) y, you need country(s) whose prefix(s) is z, e.g., SE/NEED/RTTY/BAND =(10) ZS9.

Table 4-3, continued

Common *DX PacketCluster* Commands

SE/NOA	Indicate that your computer/terminal is not ANSI-compatible.
SE/NOH	Indicate that you are not in your shack.
SE/Q x	Set your QTH as location x.
SHOW	Display requested information.
SH/A	Display names of files in archive file area.
SH/B	Display names of files in bulletin file area.
SH/C	Display physical configuration of cluster.
SH/C x	Display station connected to node whose call sign is x.
SH/CL	Display names of nodes in clusters, number of local users, number of total users and highest number of connected stations.
SH/COM	Display available Show commands.
SH/DX	Display the last five DX announcements.
SH/DX x	Display the last five DX announcements for frequency band x.
SH/DX/n	Display the last n DX announcements.
SH/DX/n x	Display the last n DX announcements for frequency band x.
SH/FI	Display names of files in general files area.
SH/FO	Display mail-forwarding database.
SH/H x	Display heading and distance to country whose prefix is x.
SH/I	Display status of inactivity function and inactivity timer value.
SH/LOC	Display your station's longitude and latitude.
SH/LOC x	Display the longitude and latitude of station whose call sign is x.
SH/LOG	Display last five entries in cluster's log.
SH/LOG n	Display last n entries in cluster's log.
SH/M x	Display MUF for country whose prefix is x.
SH/NE x	Display needed countries for station whose call sign is x.
SH/NE x	Display stations needing country whose prefix is x.
SH/NE/x	Display needed countries for mode x where x equals CW, SSB, or RTTY.

SH/NO	Display system notice.
SH/P x	Display prefix(s) starting with letter(s) x.
SH/QSL x	Display QSL information for station whose call sign is x.
SH/S x	Display sunrise and sunset times for country whose prefix is x.
SH/U	Display call signs of stations connected to the cluster.
SH/V	Display version of the cluster software.
SH/W	Display last five WWV propagation announcements.
TALK	Talk to another station.
T x	Talk to station whose call sign is x. Send <CTRL-Z> to terminate talk mode.
T x y	Send one-line message y to station whose call sign is x.
TYPE	Display a file.
TY/x y	Display file named y stored in file area named x.
TY/x/n y	Display n lines of file named y stored in file area named x.
UPDATE	Update a custom database.
UPDATE/x	Update the database named x.
UPDATE/x/ APPEND	Add text to your entry in the database named x.
UPLOAD	Upload a file.
UP x	Upload a file named x.
UP/B x	Upload a bulletin named x.
UP/F x	Upload a file named x.
WWV	Announce and log WWV propagation information.
W SF=xxx, A=yy, K=zz,a	Announce and log WWV propagation information where xxx is the solar flux, yy is the A-index, zz is the K-index and a is the forecast.

Throughout this chapter we've discussed some of the packet activities you can enjoy here on earth. But what about "the final frontier"? What am I talking about? Space, of course!

CHAPTER 5

Packet In Space

acket radio isn't just for earthbound stations! Over the past several years packet has been a major player in the Amateur Radio satellite program. With its error-free communications advantage, packet is ideal for many satellite applications.

Most packet satellites function as orbiting message carriers. They have the advantage of being able to relay messages over great distances without resorting to HF gateways. This means they aren't at the mercy of interference and unpredictable propagation.

Packet satellites also transmit telemetry you can copy at home. If you have an interest in space, it's fascinating to watch conditions onboard the spacecraft (such as solar array voltage or internal temperature). Some packet satellites are equipped with cameras that give you spectacular views of the earth below. The images are digitized and relayed to your station as streams of packet data. Let's take a look at several packet satellites and find out what you'll need to use them.

A High-Flying DOVE

DOVE, otherwise known as AMSAT-OSCAR 17, is one of the easiest packet satellites for the beginner. You can't transmit to DOVE, but you can monitor its packet telemetry *without* special equipment. All you need is the packet station

```
RAW DATA
DOVE-1>TLM:00:5A 01:5A 02:88 03:32 04:59 05:58 06:6C 07:4A 08:6C 09:68 0A:A2
0B:EC 0C:E8 0D:DC 0E:3F 0F:24 10:D8 11:93 12:00 13:D1 14:9B 15:AE
16:83 17:7C 18:76 19:7E 1A:7C 1B:45 1C:84 1D:7B 1E:C4 1F:6C 20:CF
DOVE-1>TLM:21:BB 22:79 23:26 24:22 25:26 26:01 27:04 28:02 29:3A 2A:02 2B:73
2C:01 2D:7C 2E:58 2F:A2 30:D0 31:A2 32:17 33:6B 34:AC 35:A2 36:A6
37:A8 38:86 39:A2 3A:01
DOVE-1>STATUS: 80 00 00 85 B0 18 77 02 00 B0 00 00 B0 00 00 00 00 00 00 00
DOVE-1>BCRXMT:vary= 21.375 vmax= 21.774 temp= 7.871
DOVE-1>BCRXMT:vbat= 11.539 vlo1= 10.627 vlo2= 10.127 vmax= 11.627 temp= 3.030
DOVE-1>WASH:wash addr:26c0:0000, edac=0x61
DOVE-1>TIME-1:PHT: uptime is 086/01:14:32. Time is Sat Mar 10 15:43:26 1990

DECODED TELEMETRY

DOVE   uptime is 086/01:14:32.  Time is Sat Mar 10 15:43:26 199
```

Rx E/F Audio(W):	2.21 V	Rx E/F Audio(N):	2.21 V	Mixer Bias V:	1.39 V
Osc. Bisd V:	0.51 V	Rx A Audio (W):	2.19 V	Rx A Audio (N):	2.16 V
Rx A DISC:	0.41 k	Rx A S meter:	74.00 C	Rx E/F DISC:	-1.08 k
Rx E/F S meter:	104.00 C	+5 Volt Bus:	4.94 V	+5V Rx Current:	0.02 A
+2.5V VREF:	2.51 V	8.5V BUS:	8.60 V	IR Detector:	63.00 C
LO Monitor I:	0.00 A	+10V Bus:	10.96 V	GASFET Bias I:	0.00 A
Ground REF:	0.00 V	+Z Array V:	21.38 V	Rx Temp:	7.26 D
+X (RX) temp:	-4.24 D	Bat 1 V:	1.35 V	Bat 2 V:	1.36 V
Bat 3 V:	1.38 V	Bat 4 V:	1.34 V	Bat 5 V:	1.37 V
Bat 6 V:	1.57 V	Bat 7 V:	1.36 V	Bat 8 V:	1.37 V
Array V:	21.32 V	+5V Bus:	5.30 V	+8.5V Bus:	8.85 V
+10V Bus:	11.54 V	BCR Set Point:	131.48 C	BCR Load Cur:	0.18 A
+8.5V Bus Cur:	0.06 A	+5V Bus Cur:	0.17 A	-X Array Cur:	-0.01 A
+X Array Cur:	-0.00 A	-Y Array Cur:	-0.01 A	+Y Array Cur:	0.12 A
-Z Array Cur:	-0.01 A	+Z Array Cur:	0.25 A	Ext Power Cur:	-0.02 A
BCR Input Cur:	0.45 A	BCR Output Cur:	0.29 A	Bat 1 Temp:	3.02 D
Bat 2 Temp:	-24.81 D	Baseplt Temp:	3.02 D	FM TX#1 RF OUT:	0.05 W
FM TX#2 RF OUT:	0.97 W	PSK TX HPA Temp	-3.03 D	+Y Array Temp:	3.02 D
RC PSK HPA Temp	0.60 D	RC PSK BP Temp:	-0.61 D	+Z Array Temp:	19.97 D
S band HPA Temp	8.02 D	S band TX Out:	-0.04 W		

Fig 5-1—You can easily monitor packet telemetry from the DOVE satellite in its *raw data* format using the packet station you own now. If you purchase telemetry-decoding software, you'll be able to watch the changing conditions onboard the spacecraft.

you probably own already (a TNC, a computer or terminal, a 2-meter FM transceiver and some sort of outdoor antenna).

DOVE is one of several *Microsats* presently in orbit. They're called Microsats because of their tiny size (about 9 inches on each side). DOVE's primary mission is education. It transmits streams of packet telemetry and occasional bulletins on 145.825 MHz (see Fig 5-1). Since DOVE is a *LEO* (*Low Earth Orbiting*) satellite, its signal is very easy to hear.

Set up your TNC as you would for normal operation and switch your FM transceiver to 145.825 MHz. As DOVE rises above the horizon, you'll begin to see streams of data flowing across your monitor. After you get tired of watching raw data, you'll want to find out what it means. There are several programs available to decode DOVE telemetry. See the

Packet Info Guide for more details.

Connecting to the Cosmonauts

Listening to packet satellite signals is fun, but I bet you're itching to *connect* to a spacecraft! A good candidate for your first packet connection is the Russian *Mir* space station.

Mir has been occupied by Russian cosmonauts for several years as a laboratory for testing human responses to long-duration space flights. The *Mir* studies are extremely important for future manned missions to Mars and beyond.

To combat boredom, an Amateur Radio station was installed. The cosmonauts pass amateur license tests and are assigned special *Mir* call signs (such as U9MIR) prior to launch. When they reach the station, they operate 2-meter FM voice or packet.

Like the DOVE satellite, *Mir*'s signal is powerful. You'll usually find it on 145.55 or 145.85 MHz, and you won't need sophisticated equipment to hear it—or to be heard. Just your regular packet station with an outside antenna—such as a ground plane—will do the job. Its orbit provides a couple of very good "passes" each day for most areas.

The *Mir* packet station uses standard 1200-baud AFSK packet—the same packet format you use here on earth. Once again, special equipment *is not* required to communicate with *Mir*. Most packeteers connect to *Mir's* personal packet mail-box where they leave messages for the cosmonauts (or anyone else) and pick up their replies. Others prefer to use *Mir* as an orbiting digipeater for brief connections to other stations on earth.

The biggest challenge to using *Mir* on packet is inter-ference—lots of interference! With the signal coverage the space station enjoys, you can imagine how many hams might be trying to connect to *Mir* at the same time. This creates pure chaos as far as its FM receiver is concerned.

Packet Tips: Finding the Satellites

How can you predict when a satellite is about to fly over your area? To answer that question you need to know the satellite's *orbital elements*.

An orbital element set is merely a collection of numbers that describes the movement of an object in space. By feeding the numbers to a computer program, you can determine exactly where a satellite is (or will be) at any time. When you see an element set for the first time, it will look pretty confusing. You'll see many strange terms such as *Mean anomaly, Argument of perigee* and so on. If you're curious, get a copy of the *Satellite Experimenter's Handbook* (available from your favorite dealer or the ARRL) and you'll learn all about those definitions—and more. For our purposes, however, consider the words as labels for the numbers that appear beside them.

Finding the Elements

There are several sources for orbital elements:

☐ Satellite newsletters
☐ W1AW RTTY and AMTOR bulletins
☐ Packet bulletin boards
☐ Telephone bulletin boards
☐ AMSAT nets

If you're able to connect to the mailbox, the constant bombardment of signals may make it difficult for you to post your message. (Remember that you may only have a few minutes before the space station slips below your horizon.) Here are a couple of tips to improve your chances:

☐ Listen before you start sending your connect requests. Monitor a few transmissions and make sure you have the correct call sign. The call sign changes whenever a new crew occupies the station.

(See the Packet Info Guide for more information.)

If you have an HF radio, RTTY/AMTOR capability, a packet TNC, a telephone modem or the necessary cash for a subscription, you'll always be able to get the latest orbital elements for the satellites you want to track. If all else fails, there is probably someone in your area who has access to the elements. Ask around at your next club meeting.

Using the Elements

If you have a computer in your shack, you're in luck! There are many programs on the market that will take your orbital elements and magically produce satellite schedules.

Among other things, the programs tell you when satellites will appear above your local horizon and how high they will rise in the sky (their elevation). When working satellites, the higher the elevation the better. Higher elevation means less distance between you and the satellite with less signal loss from atmospheric absorption.

Some programs also display detailed maps showing the *ground track* (the satellite's path over the ground). AMSAT (the Radio Amateur Satellite Corporation) offers satellite tracking software for a variety of computers. See the Packet Info Guide.

❏ Use as much power as you have available. If there were only a couple of stations competing for *Mir*'s receiver, you'd need only a couple of watts to have a decent chance of connecting. During a normal pass, however, there are usually *dozens* of stations blasting out connect requests. The stations that pack the bigger punches seem to win consistently.

❏ Try connecting during "unpopular" hours. If you have the stamina to sit up and wait for a late-night pass, you may have a better opportunity to make a connection.

You can recognize the mailbox by its SSID. Usually, it's -1. At the time of this writing, for example, the *Mir* mailbox was operating under the call sign RØMIR-1.

When you finally connect to the mailbox, make your message entry *short*. The station will be out of range before you know it and other hams will be waiting to try their luck. Some packet software permits the user to create a message file before attempting to connect. If your software offers this feature, it will come in handy for *Mir*.

Using Mir as a Digipeater

Using the *Mir* packet station as a digipeater is discouraged in some areas because of the interference it causes. After all, each packet that's sent to *Mir* for digipeating potentially blocks the signal of another ham who's trying to reach the mailbox. As this book went to press, the cosmonauts had not expressed an opinion for or against digipeating. Until they do, use your best judgment.

By activating the **MCOM** function in your TNC, you'll be able to watch *Mir* sending disconnect messages to eager packeteers. If you see a large number of disconnects being sent to various call signs, you can bet the *Mir* packet station is swamped. Try again on another orbit.

If you decide to send a few CQ beacons, set your **UNPROTO** parameter to **CQ VIA <call sign>**. The call sign should be that of the *Mir* digipeater, *not* the mailbox. If the mailbox call sign is RØMIR-1, for example, the digipeater is likely to be RØMIR. Although beacon packets are unconnected (you're not sending them to a specific station), you use the **UNPROTO** parameter to specify the call sign of the station that's acting as your digipeater. You can use this same technique for your terrestrial beacons, but some nodes and digipeaters may be configured to reject beacon packets.

Switch your **MCON** function **ON** before you begin

sending beacons. This allows you to watch packets from *Mir* while you're waiting for an answer. If the *Mir* digipeater hears your beacon and repeats it, you may see the text on your screen. Beacon every 30 seconds, but turn it off if you don't make a connection within 3 minutes. If another station responds to your beacon, quickly send a short message. Include your name, location and a signal report.

Hello! Name is Steve and I live in Wallingford, CT. You're 599. >>

Why do stations even bother digipeating through *Mir* if the exchanges are so brief? The answer is that many hams who relay their packets through *Mir* are looking for contacts that will count for awards—such as the ARRL's VUCC Satellite award.

Soar with SAREX!

SAREX, the *S*huttle *A*mateur *R*adio *E*xperiment, is a continuing series of Amateur Radio operations from US space shuttle missions. The first SAREX operations employed 2-meter FM voice, but more recent flights have also used packet. The variety of modes in use depends on the available cargo space. In addition, not every shuttle astronaut is a licensed ham, so not every shuttle mission has an active SAREX operation. Check *QST* for the latest news on upcoming SAREX missions.

Unlike *Mir*, SAREX usually uses a 600-kHz split-frequency scheme to accommodate standard 2-meter FM transceivers. Earthbound DXpeditions use split-frequency operation to maximize the number of stations they can work. The same is true for SAREX. With this thought in mind, you can appreciate the importance of knowing which frequencies are being used for the uplinks and downlinks. (They will be published in *QST* or transmitted by W1AW.) Whatever you do, *never* transmit on the SAREX downlink frequency. This mistake will make you very unpopular very quickly!

You'll need the shuttle's orbital elements to predict when it will be in range and these will be available through the sources we've already discussed (see the sidebar "Packet Tips: Finding the Satellites"). You can use the same 2-meter equipment for SAREX as you do for *Mir*. During one of the "packet robot" operations on a previous flight, I managed to connect using my trusty ground plane. The shuttle was only 12° above my horizon at the time, but my signal still made it!

Mail Via Satellite!

A number of satellites have the capability to store and forward packet mail and other files. As we discussed previously, packet satellites (or *Pacsats*) have the advantage of being able to forward packet messages on a timely basis regardless of band conditions. With their worldwide coverage, the Pacsats have a strong international flavor. By

Packet Tips: The SAREX Packet Robots

Several SAREX missions have used so-called "packet robots." Their purpose is to provide a means for as many packeteers as possible to contact the spacecraft. The robot listens for connect requests on designated frequencies. When a request is heard, it transmits a *contact number* and immediately disconnects. Hams who send QSL cards with the correct contact number will receive a special space shuttle QSL in return. (QSL addresses are published immediately before and after SAREX missions.)

In addition to providing quick connections for packet operators, the robot transmits a list of the call signs it's heard recently. If you see your call sign on the list, you'll know that your signal was received.

When you're attempting to connect to the space shuttle, it's a good idea to turn on your **MCOM** *and* **MCON**

connecting to these innovative satellites, you'll exchange mail with hams throughout the world. From the standpoint of ground station equipment, the current Pacsats can be separated into two categories: those operating at 1200 bits per second and those operating at 9600 bits per second.

1200 bit/s Satellites

AMSAT-OSCARs 16 and 19, Fuji-OSCAR 20 and ITAMSAT-OSCAR 26 are 1200 bit/s satellites. They also travel in low orbits. This means you won't need elaborate antennas or high RF output. Depending on what you own already, some additional equipment may be necessary (see Fig 5-2). Unlike DOVE, *Mir* or SAREX, the Pacsats transmit on one band and receive on another. For OSCARs 16, 19, 20 and 26, you can use your 2-meter FM transceiver for the uplink, but you'll need a 435-MHz SSB receiver for the downlink.

Special satellite packet modems are also required.

functions. MCOM will allow you to see the transmissions from the spacecraft. With MCON on, you'll be able to monitor this activity even while your TNC is sending connect requests.

To connect to a packet robot, just send your request in the same way you would to any station on earth. For example:

CONNECT W5RRR-1

The call sign of the packet robot changes with each mission. Watch for details in *QST* or monitor W1AW bulletins. You can also watch the robot transmissions with your MCOM function activated and note the call sign when you see it.

Packets must be transmitted to these satellites using *Manchester-encoded* frequency-shift keying (FSK). This is substantially different from the AFSK you use for packet connections here on earth. In addition, the satellites will transmit their packets to you on single sideband (SSB) using phase-shift keying (PSK). Don't worry about these unusual modulation schemes. Your satellite modem will sort them out for you.

Another nifty feature included in most 1200 bit/s satellite

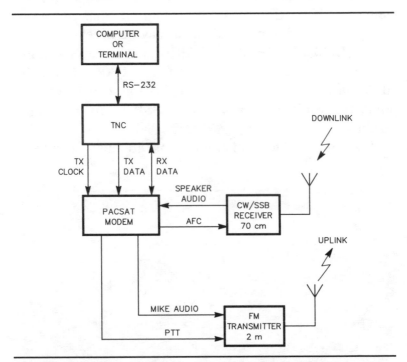

Fig 5-2—Here's a typical ground-station configuration for working the 1200-bit/s Pacsats. Your TNC is connected to a special Pacsat modem which supplies Manchester-encoded FSK for the 2-meter FM uplink. It also decodes the PSK satellite transmissions received on the 70-cm downlink. Note the AFC line from the Pacsat modem to the 70-cm receiver. The modem uses this line to tune the receiver automatically to compensate for Doppler shift on the downlink.

modems is the ability to automatically compensate for rapid downlink frequency shifting (see the sidebar "Packet Tips: The Mystery of the Shifty Signal"). If your 435-MHz receiver has some form of external frequency control (many modern transceivers feature frequency UP/DOWN buttons on the microphone), the modem will use it to keep the downlink signal tuned precisely.

Satellite modems can be added to most TNCs as outboard equipment. TNCs are also available with built-in satellite communications capability. See the Packet Info Guide for equipment sources.

Using the Pacsats is different from logging onto a terrestrial packet bulletin board. Most of the Pacsats use what

---

Packet Tips: The Mystery of the Shifty Signal

Doppler shift is caused by the difference in relative motion between you and the spacecraft. As it moves toward you, the signal frequencies in the downlink passband gradually *increase*. As it starts to move away, the frequencies *decrease*. It's the same effect you hear when a speeding railroad locomotive blares its horn. The frequency of the sound rises as the train moves toward you. As it passes your position and moves away, the frequency decreases. (Keep this in mind the next time you're stopped at a crossing. Listen to how the locomotive horn sounds *before* and *after* it reaches the crossing.)

Doppler shift is not a serious concern when working DOVE, *Mir* or SAREX. It *does* become serious when you're using the 1200 bit/s Pacsats. Because their narrow downlink signals are in the 420-MHz band, Doppler shift is very pronounced and difficult to correct manually. (The degree of Doppler shifting increases with frequency.) That's why 1200 bit/s satellite modems have an automatic frequency control feature. It tracks the frequency shift so you won't have to worry about it!

---

is known as a *broadcast protocol* to transfer information. (The only exception is Fuji-OSCAR 20.) The beauty of this system is that you don't have to actually connect to the satellite to receive a file.

When a Pacsat passes overhead, stations on the ground send their requests. Some may want to download a specific message or file. Others may simply want a listing (a directory) of what is available on the satellite. The satellite services up to 25 stations at a time by stacking their call signs in a rotating queue. As each call sign moves to the front of the line, a portion of the desired data is sent.

As a passive monitoring station (you haven't sent anything to the Pacsat), you have the opportunity to receive the same information that other stations have requested. Let's say that someone has asked for a directory. When their request reaches the front of the queue, the Pacsat starts sending their data. The target station receives the information, but so do you. As a result, your directory for that particular satellite is automatically updated—thanks to the other guy's request!

If you miss any information, your software remembers the missing elements (known as *holes*). You can request a "hole fill" from the Pacsat and it will send only the data you need to fill the gaps. The purpose of this system is to reduce the amount of communication between groundstations and satellites. Because the satellites are only available for short periods of time, they need to operate as efficiently as possible.

There are two software packages you can use to communicate with the Pacsats:

❏ *PB* allows you to receive data, request specific information, fill holes and so on. *PG* is used strictly for uploading messages and other files. Both programs are included in the *Digital Satellite Guide* available from AMSAT. See the Packet Info Guide for more information.

❏ *WISP* is a *PB/PG*-type program for Microsoft

Table 5-1

Pacsat Frequencies

Satellite	Uplink(s) (MHz)	Downlink(s) (MHz)	Data Format	
AMSAT-OSCAR 16	145.90 145.92 145.94 145.96	437.025 437.05 2401.10	Uplink: Downlink:	1200 bit/s FSK 1200 bit/s PSK
DOVE-OSCAR 17	None	145.825	Telemetry at 1200 bit/s AFSK (FM)	
WEBERSAT-OSCAR 18	None	437.10	1200 bit/s PSK digitized images	
LUSAT-OSCAR 19	145.84 145.86 145.88 145.90	437.125 437.15	Uplink: Downlink:	1200 bit/s FSK 1200 bit/s PSK
Fuji-OSCAR 20	145.85 145.87 145.89 145.91	435.910	Uplink: Downlink:	1200 bit/s FSK 1200 bit/s PSK
UoSAT-OSCAR 22	145.90 145.975	435.120	Uplink: Downlink:	9600 bit/s FSK 9600 bit/s FSK
KITSAT-OSCAR 23	145.85 145.90	435.175	Uplink: Downlink:	9600 bit/s FSK 9600 bit/s FSK
KITSAT-OSCAR 25	145.87 145.98	436.50	Uplink: Downlink:	9600 bit/s FSK 9600 bit/s FSK
ITAMSAT-OSCAR 26	145.875 145.900 145.925 145.950	435.87	Uplink:	1200 bit/s FSK 1200 bit/s PSK

Windows. It was developed by Chris Jackson, ZL2TPO. *WISP* is also available from AMSAT.

By using *PB/PG* or *WISP*, some Pacsat users have been

able to fully automate their satellite stations. While they're at work or school, their stations are on alert, ready to download information as the satellites streak by. When they come home, they're greeted by hundreds of kilobytes of information—everything from messages to images and software!

9600 bit/s Satellites

The latest trend in packet satellites is to use higher data rates—typically 9600 bit/s. The motivation behind this trend is obvious: If you can only access a satellite for 10 or 15 minutes, you want to send and receive data as rapidly as possible.

The primary 9600-bit/s store-and-forward satellites are UoSAT-OSCAR 22, KITSAT-OSCAR 23 and KITSAT-OSCAR 25. Although the data rate is much higher than other packet satellites, these satellites use FM audio frequency shift keying (AFSK) on the uplinks and FM FSK on the downlinks. *PB/PG* or *WISP* software is also used.

You'll need 9600-bit/s satellite modems to access these powerful birds. Transmitting and receiving 9600-bit/s packet signals can be a bit tricky. Standard 2-meter FM transmitters may not pass the audio tones without modification (injecting the audio *after* the microphone amplifier stages, for example). Receiver IF filters may have to be changed to 20 or 30 kHz. The receive audio must be tapped immediately after the discriminator since the audio amplifier circuits will seriously distort the tones. An alternative is to purchase packet transceivers specifically designed to handle higher data rates (see the Packet Info Guide). As these satellites become more popular, you can expect to see more 9600-bit/s gear on the market.

Satellite Gateways

What if you want to send messages to hams in distant states or other countries, but you lack the equipment to reach the Pacsats? There *is* an alternative: packet satellite gateways.

Satellite gateways are like multi-port PBBSs, except that one of their ports is a Pacsat uplink/downlink! A list of gateway stations is shown in Table 5-2. The trick is to address your message in the proper manner. Let's say that I wanted to send a message to VK3ANQ near Melbourne, Australia. Here's how I would address it:

SP REQSAT @ WA0PTV

I'm sending this message to my nearest satellite gateway station. Sending it to "REQSAT" means that I want it to be processed by the REQSAT gateway software. When the PBBS asks for the subject, I can enter whatever comes to mind.

Table 5-2

United States Satellite Gateways

PBBS	Location	Hierarchical Address
AA6QD	Los Osos, CA	CA.USA.NA
KL7AA	Anchorage, AK	AK.USA.NA
WA0PTV	Fredonia, NY	NY.USA.NA
KF4WQ	Selinsgrove, NC	NC.USA.NA
K0PFX	Manchester, MO	MO.USA.NA
WV9O	La Porte, IN	IN.USA.NA
NR3U	Selinsgrove, PA	PA.USA.NA
N0GIB	Sioux Falls, SD	SD.USA.NA
WB5EKW	Abilene	TX.USA.NA
WH6AQ	Honolulu, HI	HI.USA.OC
W7LUS	Ft Lauderdale, FL	FL.USA.NA
K7PYK	Phoenix, AZ	AZ.USA.NA

How are things down under?

Now the PBBS will prompt me to begin entering the text of the message. The *first line* of the message text must be the actual packet address of the destination station. Include the hierarchical address if you know it.

SP VK3ANQ @ VK3PPP.VIC.AUS.OC

It looks like I'm repeating the **SP** command doesn't it? Don't worry; your PBBS will treat it as text. When the satellite delivers the message to the gateway station in Australia, however, this is the command and address format used to send the message to the target PBBS.

Simply enter the rest of your message on the following lines. Keep it very short since satellite storage capacity is limited. End your message with the usual CONTROL-Z or /EX and you're done. The message will be forwarded to the gateway PBBS and then relayed to its destination via satellite!

Not all satellite gateways use the REQSAT format. Before attempting to send a message, contact the SysOp at your nearest gateway and ask which format he or she uses. New gateway stations are appearing and old ones are disappearing on a regular basis. Monitor your PBBS for bulletins concerning changes in the satellite gateway system.

Eyes In The Sky

Few of us will ever get the chance to journey into space. Even so, we can still get a glimpse of what it's like to be an astronaut. By using your packet station equipment, you can receive digitized *images* from cameras onboard AMSAT-OSCAR 18 (otherwise known as WEBERSAT), UoSAT-OSCAR 22, KITSAT-OSCAR 23 or KITSAT-OSCAR 25.

OSCAR 18 is another member of the Microsat family launched in 1990. It carries a miniature color TV camera and

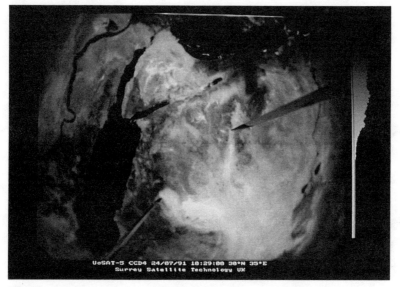

UoSAT-5 CCD4 24/07/91 18:29:00 30°N 35°E
Surrey Satellite Technology UK

The Red Sea as seen by UoSAT-OSCAR 22. This image was received by amateurs at the University of Surrey in the United Kingdom.

transmits one or two images per week. The image is converted to packet data and sent at 1200 bit/s using PSK. Special software (known as *Weberware*) is required to decode and view the images on your computer's CGA, EGA or VGA monitor. *Weberware* is available from AMSAT (see the Packet Info Guide). You don't need to connect to OSCAR 18 to receive the images. Just monitor its downlink frequency: 437.100 MHz.

UoSAT-OSCAR 22 provides superb pictures from its onboard camera. The images are transmitted using FM FSK at 9600 bit/s on 435.120 MHz. *PB/PG* or *WISP* software and 9600 bit/s modems are required to connect to the satellite and download the image files. A program known as *DISPLAY4* is used to decode the image data (see the Packet Info Guide). You'll also see fascinating imagery from OSCARs 23 and 25.

CHAPTER 6

Different Approaches to Packet

hroughout this book we've discussed terrestrial packet in its most popular form: AX.25 protocol using *NET/ROM* node networks and standard TNCs. Hams are never content with the status quo, however. If it's possible to improve a process or design, you can bet an Amateur Radio operator is working on it!

Where packet radio is concerned, this inventive obsession has created a number of alternative approaches. Some are supported enthusiastically by packet-active hams in many areas of the country. They have the potential for worldwide acceptance and could very well be the wave of the future. Take a look at several of the better-known alternatives and decide for yourself.

A ROSE is a ROSE is a ROSE

Several years ago, the Radio Amateur Telecommunications Society (RATS) developed a networking protocol known as *RATS* *O*pen *S*ystem *E*nvironment, or *ROSE*. Like networks based on *NET/ROM* nodes, the objective of ROSE is to let the network do the work when you're trying to connect to another station.

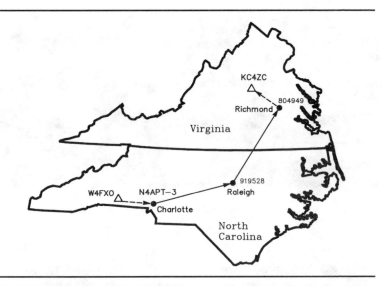

Fig 6-1—In this example, W4FXO, near Charlotte, North Carolina, uses the ROSE network to establish a connection to KC4ZC northwest of Richmond, Virginia. All that W4FXO has to do is issue a connect request that includes his local ROSE switch (N4APT-3) and the ROSE address of the switch nearest KC4ZC (804949). As soon as the request is sent, the network takes over. In this example, the connection to KC4ZC is established by using a ROSE switch in Raleigh.

Do you remember how to use a *NET/ROM* node network to connect to a distant station? You check a nearby node to see if the call sign of the node you want to reach appears on its *node list*. If it does, you "ask" the node to connect to it. The node will attempt to fulfill your request, using whatever frequencies and additional nodes are necessary. Once you connect to the target node, you send a connect request to the desired station.

ROSE provides a less complicated, and more reliable, way to make a connection. Using a ROSE network is similar to using the telephone. ROSE nodes are frequently referred to as *switches*, and each switch has its own address based on the telephone area code and the first 3 digits of the local exchange. A ROSE switch in my area of Connecticut, for example, may have an address of 203555. 203 is the area code and 555 is the local telephone exchange.

The ROSE network uses this addressing system to create reliable routes for packets. It also makes it very easy for users to establish connections. Let's say you wanted to connect to WB8ISZ. WB8QVC is the call sign of your local ROSE switch and 513628 is the address of the switch nearest WB8ISZ. Here's how you'd do it:

CONNECT WB8ISZ VIA WB8QVC-3,513628

The call sign of your local ROSE switch immediately follows the "via," followed by the ROSE address for the switch nearest WB8ISZ. (Notice that the call signs of ROSE switches have an SSID of -3.)

There's no guesswork on your part. The ROSE network will attempt to create a path between you and WB8ISZ automatically. The network will acknowledge your request by sending:

*** CONNECTED to WB8ISZ VIA WB8QVC-3,513628

Call being setup

Although the message indicates that you're connected to WB8ISZ, the network will send another message after the connection has actually been established:

Call Complete to WB8ISZ @ 3100513628

Why does the ROSE address suddenly look different? It was 513628, but now it's 3100513628! No need to worry. The network is just showing you the complete ROSE address of the switch that made the final link. "3100" is a *country* identification. The Amateur Radio Data Network Identification code for the United States is 3100. This is added to the address to create 3100513628.

At this point you're connected to WB8ISZ through the ROSE network. The path between your stations will be the most reliable path the network can find. Every ROSE switch in the path "knows" the proper routing for all the packets sent between you and WB8ISZ.

When you finish your conversation, you simply disconnect as you would on a *NET/ROM* system. As the link is broken, each switch in the path will erase your route from its memory. As a result, this is what you'll see:

*** Call Clearing
*** 0000 3100513628 Remote Station cleared connection

Unless you wish to set up a ROSE switch of your own, you won't need special equipment or software to use the network. You can access a ROSE network today if a switch is available in your area. As we just discussed, all you need to know is the call sign of your local switch and the ROSE address of the switch nearest to any stations you want to contact (individuals or PBBSs).

ROSE networks are appearing in many areas of the country. They are especially popular in the southeast and mid-Atlantic states. ROSE addresses and system maps are available from RATS, PO Box 93, Park Ridge, NJ 07656-0093. Send a business-sized SASE with your request.

Networking the TexNet Way

TexNet is a high speed, centralized packet networking system developed by the Texas Packet Radio Society (TPRS). Designed for local and regional use, TexNet provides AX.25-compatible access on the 2-meter band at 1200 bit/s. This allows packeteers to use TexNet without investing in additional equipment or software. The node-to-node backbones operate in the 70-cm band with data moving through the network at 9600 bit/s. Telephone links are also used to bridge some gaps in the system.

The network offers a number of services to its users. Two conference levels are available by simply connecting to the proper node according to its SSID. By connecting to W5YR-2, for example, you'll join the first conference level. Connecting to W5YR-3 places you in the second level. When you connect to a conference, you can chat with anyone else on the network in roundtable fashion.

When you want to connect to a specific station, you access the network by connecting to a node with an SSID of -4. Once you're connected, issue the connect request to the desired station (with your TNC in the *converse* mode, of course) and include the alias of its local TexNet node:

Connect WD5IVD @ DALLAS

TexNet will send your request to the DALLAS node on 70 cm at 9600 bit/s. You won't have to wait very long to find out if you're successful!

Every TexNet network is served by a single *Packet Message System*, or *PMS*. By using only one PMS, the network isn't bogged down with constant mail forwarding. Even if you're some distance from the PMS, with the speed and efficiency of TexNet you'll hardly notice the delay. Just connect to the network node (any TexNet node with an SSID of -4) and send the **Message** command. Within seconds you'll be connected to the TexNet PMS. Many TexNet PMSs offer sophisticated features such as up-to-the-minute weather data. To see this information you only have to send the **Weather** command.

The prime movers of TexNet are George Baker, W5YR, Tom McDermott, N5EG, and Tom Aschenbrenner, WB5PUC. TexNet presently connects the Texas cities of Austin, Corpus Christi, Dallas, Houston, Midland, Plano,

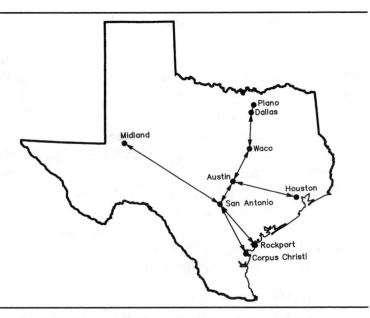

Fig 6-2—TexNet has created an efficient packet network for amateurs throughout Texas. Users access their local TexNet nodes on 2 meters. Data is shuttled through the network at 9600 bit/s using 70-cm backbone links and telephone lines. TexNet has been so successful, it has appeared in other states such as Oklahoma and Michigan, Arkansas, Indiana and New Mexico.

Rockport, San Antonio and Waco. Despite the size of this system, users on the longest path (between Midland and Rockport) experience only a 6-second round-trip delay when exchanging packet data. TexNet networks have also appeared in Oklahoma, Michigan, Arkansas, Indiana and New Mexico.

Versatile KA-Nodes

Most TNCs and MCPs manufactured by Kantronics Incorporated include a KA-Node. As a result, you'll see KA-Nodes throughout the country.

A KA-Node is very similar to a *NET/ROM* node. Once a KA-Node acknowledges error-free reception of your packet, it will pass it along to its destination without additional acknowledgments from your station. Unlike *NET/ROM* nodes, however, a KA-Node cannot perform automatic routing. You can't connect to a single KA-Node and ask it to make a multi-node connection to a distant station. Instead, you have to literally jump from one node to another, sending separate connect requests as you go. In this sense, a KA-Node is similar to a traditional digipeater.

Because of the sheer number of KA-Nodes in the packet community, you should familiarize yourself with their operation. You never know when you'll need to use a KA-Node to make a connection!

You'll find that KA-Nodes are often operated erratically. Some individuals and groups keep their KA-Nodes on the air at all times, but many others activate their nodes only when they're at home. (The KA-Node you used last night may not be available tomorrow night!) It pays to keep a log of the KA-Nodes in your area. Drop a message to the operator and ask if he or she keeps a regular schedule.

Let's say you discover the WDØEMR-2 KA-Node on the 2-meter band. Connect to the node and you'll see a response similar to the one shown below.

*** CONNECTED to WDØEMR-2

CONNECTED TO NODE WDØEMR-2 CHANNEL A ENTER COMMAND B,C,J,N,X OR HELP ?

The commands are straightforward. **B** means "Bye," or disconnect and **C** stands for "Connect." Sending a **J** will download the "JHeard," or "Just Heard" stations list. **N** displays a list of local KA-Nodes and *NET/ROM* nodes. **X** is the cross-connect or gateway function (KAMs and KPC-4s only).

If you send the **J** command you'll see a list of station call

---

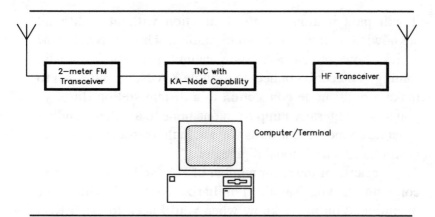

Fig 6-3—Full-featured KA-Nodes can provide crossband links between the VHF and HF bands. In this example, 2-meter users use the KA-Node to access the HF bands for long-distance packet contacts. HF packeteers can also use KA-Nodes to reach local VHF packet networks.

signs and/or aliases. If you are connected to a KA-Node installed in a KAM, the list may look like this:

WKØM-3/V	**01/08/92**	**09:25:15**
NFØT/V	**01/08/92**	**09:30:00**
SPRING/V	**01/08/92**	**09:41:35**
SM5DFH/H	**01/08/92**	**09:43:05**

The /V signifies that the station was heard on the KAM's VHF port. By using the Connect command, you can connect to any of these stations if they are still on the air. The /H means that the station was heard on the HF port.

As you scanned the list, did you notice that the KA-Node heard SM5DFH on its HF port? Assuming that little time has passed, the SM5 station may still be on the frequency. If you want to try your luck, use the cross-connect command:

XC SM5DFH

The KA-Node will attempt to establish a connection to

SM5DFH on its HF port. In other words, it will pass your connect request to SM5DFH using HF equipment. If the attempt is successful, you'll see:

LINK MADE
CONNECTED TO SM5DFH

As you send your packets, the KA-Node will receive them on 2 meters and relay to SM5DFH on the HF bands. It's important to note that this method of cross connecting works both ways. If SM5DFH discovered the KA-Node on HF, he could use it to explore activity in your local area. So if you have a cross-linked KA-Node nearby, don't be surprised if you get a connect request from a DX station!

TCP/IP: Pings, POPs and KISSes

If you're an active packeteer, sooner or later someone will bring up the subject of TCP/IP—Transmission Control Protocol/Internet Protocol. Of all the packet networking alternatives discussed so far, TCP/IP is the most popular.

Despite its name, TCP/IP is more than two protocols; it's actually a set of several protocols. Together they provide a high level of flexible, "intelligent" packet networking. At the time of this writing, TCP/IP networks are local and regional in nature. For long-distance mail handling, TCP/IP still relies on traditional AX.25 *NET/ROM* networks. Even so, TCP/IP enthusiasts see a future when the entire nation, and perhaps the world, will be linked by high-speed TCP/IP systems using terrestrial microwave and satellites.

Sending Mail with TCP/IP

With TCP/IP you can send local and regional mail efficiently and reliably. You need only prepare the message and leave it in your own TCP/IP "mailbox." Your computer will attempt to make a connection to the target station and deliver the message directly. There are no PBBSs involved. The message packets simply travel through the network until they reach the other station. While this is taking place, you can talk to, or receive mail from, another station. If you're fortunate enough to own *Windows*, *DESQview* or similar multitasking software, you can leave the TCP/IP program entirely and play a game or write a letter while your message is being delivered!

Before sending a message to another station, it's possible to use TCP/IP to see if the station is actually on the air. You use the **ping** function, which stands for Packet Internet Groper—no kidding! Think of it as the sonar pings used by ships and submarines to find out what's in the water around them. It works in nearly the same manner, but it's much more specific! If I want to find out if WS1O is available on the network, I just **ping** him as follows:

ping ws1o

Note that TCP/IP is *case sensitive*. That is, the use of upper- or lower-case letters is important. Many TCP/IP commands are lower-case.

If WS1O has his station on the air, I'll see:

44.88.0.23: rtt 7000 ms

This cryptic line merely confirms that WS1O is on the air. His TCP/IP address is shown (44.88.0.23) along with the time it took to send the ping from my station to his (7000 ms or 7 seconds). It lets me know that the mail I send now will arrive at his station within a reasonably short period of time.

If you want to send mail to hams in cities outside your TCP/IP network, you'll need to post it on an AX.25 bulletin board. Some AX.25 PBBSs feature special ports that allow TCP/IP users to connect to the system. If your local PBBSs doesn't have a TCP/IP port, don't worry. Your TCP/IP software provides a means to communicate with AX.25

stations (and vice versa). As TCP/IP networks expand, you'll be able to reach hams in distant cities without having to resort to the AX.25 mail forwarding system.

What if someone tries to send mail to you when you're not on the air? If they can't connect to your station, their computer will hold the message and try again later. Many TCP/IP packeteers also act as *Post Office Protocol* (POP) servers—temporary warehouses for incoming mail. When POP "clients" activate their stations, the TCP/IP software sends inquiries to their servers to see if any mail is waiting. If mail is available, it's transferred automatically!

Shooting the Breeze with TCP/IP

Talking to another station on a *NET/ROM* network can be a difficult proposition—especially if the station is distant. You can only hope that all the nodes in the path are able to relay the packets back and forth. If one of the nodes becomes unusually busy, your link to the other station could collapse. Even when the path is maintained, your packets are in direct competition with all the other packets on the network. With randomly calculated transmission delays, collisions are inevitable. As a result, the network bogs down, slowing data throughput for everyone.

TCP/IP has a unique solution for busy networks. Rather than transmitting packets at randomly determined intervals, TCP/IP stations automatically *adapt* to network delays as they occur! As network throughput slows down, active TCP/IP stations sense the change and lengthen their transmission delays accordingly. As the network speeds up, the TCP/IP stations shorten their delays to match the pace. This kind of intelligent network sharing guarantees that all packets will reach their destinations with the greatest efficiency the network can provide.

With TCP/IP's adaptive networking scheme, you can use

the *Telnet* function to chat with a ham in a distant city and rest assured that you're not overburdening the system. Your packets simply join the constantly moving "freeway" of data. They might slow down in heavy traffic, but they *will* reach their destination eventually. (This adaptive system is used for *all* TCP/IP packets, no matter what they contain.) If you want to check the status of your target station—and the path between—just use the **ping** function as we discussed earlier.

Swapping Files via TCP/IP

TCP/IP really shines when it comes to transferring files from one station to another. By using TCP/IP, you can connect to another station and transfer computer files— including software. As you can probably guess, transferring large files can take time. With TCP/IP, however, you can still send and receive mail or talk to another ham *while* the transfer is taking place!

To transfer a file from one station to another, you use the *File Transfer Protocol*, or FTP. We already know that WS1O is on the air. Let's set up an FTP link with his computer and grab a file.

ftp ws1o

When the link is established, my terminal displays:

Established
**220 ws1o.ampr.org FTP version 890421.1e ready at Sat
 Aug 13 18:22:26 19**

Now I'm ready to log in. At the command prompt I send: **USER anonymous**. My terminal displays WS1O's response.

331 Enter PASS command

No problem. Most systems allow you to use your call sign as the password.

PASS wb8imy

When I see **230 Logged in**, it's time to check his computer to see what he has to offer. All I have to do is send **dir** and my screen displays:

200 Port command okay
150 Opening data connection for LIST/public

switch.map	**1,500**	**19:57**	**06/19/92**
tcp/ip.doc	**10,000**	**02:30**	**07/01/92**
space.exe	**20,000**	**22:25**	**07/16/92**

3 files 13,617,152 bytes free. Disk size 33,400,832 bytes
Get complete, 200 bytes received

Whew! It looks complicated at first glance, doesn't it? All it's telling you, though, is that WS1O has three files available for transfer. The name of each file is shown along with its size and the date it was placed on his disk. If you have some experience with computers, this may look familiar. (You've just issued the "directory" command that's common to most machines.)

I happen to know that "space.exe" is a game that WS1O has written for my particular computer. I can transfer a copy by simply using the **get** command.

get space.exe

In response, I'll see:

200 Port command okay
150 Opening data connection for RETR /public/space.exe

As I've already pointed out, the transfer may take several minutes or several hours, depending on the size of the file and the level of activity on the network. That's okay because I can do something else (mow the lawn!) while my computer handles the transfer. When the transfer is complete, my screen displays:

Get complete, 20,000 bytes received
226 File sent ok

Great! The transfer was a success. I can send a file to his station using the **put** command, or I can **quit** and go play the game!

221 Goodbye!
FTP session 1 closed: EOF

TCP/IP Switches

Most TCP/IP networks depend on dedicated switches to move data through the system. *NET/ROM* nodes are also used when necessary. Like *NET/ROM* nodes, TCP/IP switches communicate with each other over high-speed backbone links on the 222- or 420-MHz bands. Many TCP/IP packeteers access their local switches on the 2-meter band with 144.91 MHz being a popular frequency.

When you try to contact another station using TCP/IP, all network routing is performed automatically according to the TCP/IP address of the station you're trying to reach. In fact, TCP/IP networks are transparent to the average user. Your packets may flow back and forth through several switches and you won't even know it!

On conventional *NET/ROM* networks, access to backbone links is restricted. This isn't true on TCP/IP. Not only are you allowed to use the backbones, you're actually *encouraged* to do so! If you have the necessary equipment to communicate at the proper frequencies and data rates, you can tap into the high-speed TCP/IP backbones directly. By doing so, you'll be able to handle data at much higher rates. This benefits you and everyone else on the network.

What Do I Need to Run TCP/IP?

You'll be pleased to know that you can join the TCP/IP

community with the packet equipment you probably own already. In terms of hardware, all you need is a computer (it must be a computer, not a terminal), a 2-meter FM transceiver and a TNC with *KISS* capability.

What in the world is KISS? KISS stands for Keep It Simple, Stupid (don't you love these terms?) and most modern TNCs and MCPs have this feature. When you place your TNC in the KISS mode, you disable the AX.25 protocols and reduce the unit to a basic packet modem. All of the incoming and outgoing data will be processed directly by your computer, not your TNC.

As you might guess, the heart of your TCP/IP setup is software. The original TCP/IP software set was written by Phil Karn, KA9Q, and is called *NOSNET* or just *NOS*. The program is available for IBM-PCs and compatibles, Apple Macintoshes, Atari STs and Commodore Amigas (see the Packet Info Guide for sources). NOS takes care of all TCP/IP functions, using your "KISSable" TNC to communicate with the outside world.

The only other item you need is your own IP address. IP Address Coordinators assign addresses to new TCP/IP users. All you have to do is contact the coordinator in your area (see the Packet Info Guide).

It sounds almost too simple, doesn't it? Well, learning a new protocol such as TCP/IP presents a challenge to some. There's a completely different set of abbreviations and "buzz words" that you'll need to know. It also takes some time to completely set up the NOS software and get it running properly. Most experienced TCP/IP users are more than glad to help you solve any problems.

If you're curious, give TCP/IP a try. Since you'll still be able to connect with the AX.25 network as you did before, you have nothing to lose and everything to gain.

Digital Signal Processing

In this chapter our discussion has focused on new approaches to packet networking. How about a new approach to packet hardware? Will anything replace the standard packet TNC?

Digital signal processing (DSP) has been used by commercial communications companies and the military for many years. Within the last few years, however, DSP technology has become affordable for Amateur Radio use. Digital signal processing is more than just a fad or gimmick. The implications for the future of Amateur Radio are profound.

Robots and Fruit Cakes

As wonderful as a standard TNC or MCP may be, its major weakness is its lack of flexibility. If you have read Chapter 5, for example, you know you must add a special outboard modem to access most packet satellites. This is because your TNC is designed only for a very specific mode of operation. It's like a robot programmed to perform a single task on an assembly line. A robot that welds car bodies can't be put to work packaging fruit cakes!

MCPs are designed to address this problem. They include several digital modes *in addition* to packet, but they have their limitations, too. If a mode is incompatible with the filters and other hardware in the MCP, you're out of luck (most MCPs can't communicate with packet satellites, for example). What's needed is a data communications processor that can easily adapt to change. Putting it another way, you need a welding robot that can package fruit cakes at a moment's notice!

Digital signal processing gives you this kind of flexibility. A DSP data controller takes the incoming signal, regardless of mode, and turns it into digital data. The data is analyzed by a high-speed microprocessor and the decoding

AEA's DSP-2232 incorporates digital signal processing in its design. DSP equipment features extraordinary flexibility, allowing you to communicate using *any* Amateur Radio digital mode!

process is controlled by *software*, not hardware.

When you want to use your computer to write a letter, you select the proper program. When you're tired of writing and you want to play a game, you select a different program. The computer hasn't changed—only the software, right? The same idea applies to a DSP modem. If you're communicating with your local bulletin board, a DSP modem uses AX.25 packet processing software. If you decide that you'd like to access a Pacsat, it uses packet satellite software.

A DSP modem can be anything you want it to be. A single device can be used for terrestrial packet, satellite packet, RTTY, AMTOR, slow-scan television, weather fax and so on. If a new digital communications mode comes along, you only have to plug in the new software on a chip. It's even possible to upgrade a DSP modem by downloading software from a telephone BBS!

As long as the proper processing software is available, a DSP modem will never be obsolete. After you make your initial purchase, there's nothing else to buy except software upgrades to accommodate new modes. Several DSP units are on the market at the time of this writing. More will become available soon. Digital signal processing is also being put to use in precision audio filters and other equipment. You can count on hearing a lot more about DSP as time goes on.

A Packet-Speak Glossary

Terms in italics appear elsewhere in this Glossary.

Ack: an abbreviation for "acknowledgment." Packet stations exchange Acks to verify that data has been received without errors.

AFSK: an abbreviation for "audio frequency-shift keying." A method of digital transmission that is accomplished by varying the frequency of an audio tone. Widely used for terrestrial packet.

alias: an alternative method of addressing a packet station. For example, the call sign of my personal mailbox could be WB8IMY-4, but its alias could be STEVEBOX. Either the call sign or the alias can be used to establish a connection.

AMSAT: the Radio Amateur Satellite Corporation. An organization dedicated to designing and building Amateur Radio satellites.

AMTOR: an acronym for Amateur Teleprinting Over Radio. A popular method of error-free communication on the HF bands using an acknowledgment/ nonacknowledgment *protocol* similar to packet.

ASCII: an acronym for American Standard Code for Information Interchange (usually pronounced *as-key*). A standard method of encoding data so it can be understood by many different computers.

autobaud: a routine used by many *TNCs* and *MCPs* to automatically adapt to the data rate of a computer or *terminal*.

AX.25: the Amateur Radio version of the CCITT X.25 packet *protocol* used for computer communications over telephone lines.

backbone: a packet radio network that transfers data from one station to another. Most backbones operate at high data rates on the 222- and 420-MHz bands.

BayCom: a *terminal* and *TNC* emulation software package for IBM PCs and compatibles.

beacon: a *TNC* function that allows stations to send *unconnected packets* automatically at regular intervals. Often used for calling CQ.

bit/s: an abbreviation for "bits per second," a measurement of the rate at which data is transferred from one device to another.

bulletins: messages of general interest sent from one station to many stations throughout the packet network.

capture: to save incoming data to a disk file for later use.

CD-ROM: an acronym for Compact Disc Read Only Memory. Large amounts of data can be stored permanently on a compact disc and read very rapidly by a computer. Many sophisticated PBBSs use CD-ROMs for call sign directories and magazine bibliographies.

chat node: a packet *node* designed to allow many users to talk to each other simultaneously.

conference: a group discussion area provided on packet *nodes* or *PBBSs*.

connect: to establish a communications link between two packet radio stations.

converse: a *TNC* mode where all information entered at the computer or *terminal* is sent directly over the air.

data terminal: a device that allows a human operator to communicate with a computer or *TNC*.

DCD: an abbreviation for "data carrier detect." Many *TNCs* and *MCPs* have a DCD indicator which glows when packet data is received.

DIGICOM>64: a *terminal* and *TNC* emulation software package designed for Commodore-64 computers.

digipeater: digital repeater, a device that receives, temporarily stores and then retransmits packet radio transmissions that are specifically directed to it.

digital signal processing: using software rather than hardware to encode or decode digital signals for various modes.

disconnect: to terminate a communications link between two stations.

Doppler shift: in amateur satellite communications, a shift in the frequency of a signal caused by differences in relative motion between the satellite and the ground station.

downlink: in amateur satellite communications, the frequency a satellite uses to transmit information to a ground station.

download: the act of requesting and receiving specific data from another station.

DSP: an abbreviation for *digital signal processing*.

dumb terminal: a basic *data terminal* that provides only input and output functions. It cannot store or process data.

EIA-232-E: the current designation for interfacing computers and/or *terminals* to other devices such as *TNCs* and *MCPs*. (formerly *RS-232-C*)

fax: an abbreviation for facsimile. A digital communications mode used for transmitting high-resolution still images.

firmware: software stored in an integrated circuit memory chip.

floppy disk: removable magnetic disks used to store digital data.

flow control: stopping and restarting the transfer of data between a computer/*terminal* and a *TNC*.

forwarding: the act of relaying data from one packet station to another.

FSK: an abbreviation for "frequency shift keying."

FTP: an abbreviation for File Transport Protocol. Used by *TCP/IP* stations to transmit files.

gateway: a *node* or *PBBS* function that allows stations on different frequencies to communicate with each other.

ground track: the path of a satellite over the ground.

hacker: a computer hobbyist

hard disk: nonremovable magnetic disks used to store large amounts of data.

hierarchical address: the address of a packet station that specifies its location in ascending geographical order. (WB8IMY.#CCT.CT.USA.NA = WB8IMY in central Connecticut, Connecticut, United States of America, North America)

KaNode: a *NET/ROM*-like *node* function included in a number of *TNCs* and *MCPs* manufactured by the Kantronics Corporation.

Keplerian elements: a set of numbers used to describe the motion of an object in space (also known as *orbital elements*)

KISS: Keep It Simple, Stupid. A nonprotocol mode included in all modern *TNCs* and *MCPs*.

LEO: an acronym for Low Earth Orbiting satellite

Manchester encoding: a means of encoding data on an FM *uplink* signal for transmission to a packet satellite.

MCP: an abbreviation for Multimode Communications Processor. A digital device designed to provide packet, CW, RTTY, fax and AMTOR communications as well as other modes. Also known as a multimode data controller.

Microsats: A group of small digital communications satellites designed by AMSAT.

NET/ROM: *node* networking software designed by Ron Raikes, WA8DED, and Mike Busch, W6IXU. *NET/ ROM* nodes relay packet data from station to station without the need for end-to-end acknowledgments. When a packet is received by a node, the node assumes responsibility for relaying it to its destination. The ACK is sent only to the previous node in the path, not all the way back to the originating station.

node: a junction point in a packet network where data is relayed to other destinations. A node can support more than one user at a time and can operate on several different frequencies simultaneously.

NOS: Amateur Radio *TCP/IP* software written by Phil Karn, KA9Q.

NTS: National Traffic System. An ARRL-sponsored system for relaying messages throughout the nation and the world. NTS is supported by packet networks as well as CW, phone, RTTY and AMTOR nets.

orbital elements: a set of numbers used to describe the motion of an object in space (also known as *Keplerian elements*)

***PacketCluster*:** a network of specialized *nodes* dedicated to DX hunting and contest activities. Packeteers can check into *PacketCluster* systems and determine which DX stations are on the air and what frequencies they're using. Other information is available as well. (Also see *spot.*)

Pacsats: Amateur Radio satellites featuring packet communications capability.

PBBS: an abbreviation for Packet Bulletin Board System. A repository for packet mail, bulletins and other information within a local packet network.

ping: A *TCP/IP* function used to test path quality between one station and another.

POP: an abbreviation for Post Office Protocol. A *TCP/IP* system where one station acts as a collection point for incoming mail.

port: a circuit that allows one device (such as a computer) to communicate with another device (such as a *TNC*). In packet terminology, a port can also refer to one of several frequencies used by a *PBBS*, *node* or *gateway*.

protocol: a set of recognized procedures. AX.25 is an example of a digital packet communications protocol.

PSK: an abbreviation for Phase Shift Keying. A method of digital transmission that is accomplished by varying the phase of a carrier between two values. Often used on Pacsat *downlink* signals.

PTT: an abbreviation for Push To Talk. On a transceiver microphone, the button that's used to key the transmitter. Packet *TNCs* use the PTT line to key transmitters automatically.

RAM: an acronym for Random Access Memory. A data storage device that can be written to and read from. Commonly used to refer to memory chips within a computer or other microprocessor-controlled device.

retry: the retransmission of a packet that has been lost or received with errors.

ROSE: an acronym for RATS Open Systems Environment. A network system utilizing *node* addressing based on telephone area codes and local exchanges.

RS-232-C: The former standard for interfacing computers and/or *terminals* to other devices such as *TNCs* and *MCPs*. It was replaced by EIA-232-E, although many still use the term RS-232-C when referring to a "standard" interface.

RTTY: an acronym for Radioteletype (usually pronounced *ritty*). One of the oldest methods of digital communications. RTTY is still popular on the HF bands.

SAREX: an acronym for Shuttle Amateur Radio Experiment. Amateur Radio operations aboard US space shuttle missions.

serial port: A circuit that permits one device to communicate with another by sending information bit-by-bit. Computers and *data terminals* often communicate with *TNCs* and *MCPs* through their serial ports.

spot: DX information posted on a *PacketCluster*. Usually includes a call sign, frequency and possibly a short comment.

SSID: an abbreviation for Secondary Station Identifier. A number that is used to differentiate between two or more station functions operating under the same call sign. For example, WB8IMY-2 may be a node while WB8IMY-4 may be a PBBS.

STA: an abbreviation for Status. An indicator featured on most *TNCs* and *MCPs*. It glows whenever packets are awaiting transmission, or have been transmitted and are awaiting acknowledgment.

store and forward: The act of receiving data (messages or bulletins) and then storing them temporarily until they can be passed along to the next station.

stream switching: Using a *TNC* to conduct several packet connections simultaneously. A keyboard character is often used to switch from one stream (connection) to another.

switch: Any device that's used to relay packet data within a network. A *node, digipeater* or *PBBS* can function as a switch.

SysOp: an abbreviation for System Operator. The individual who operates and maintains a *PBBS*.

TCP/IP: an abbreviation for Transmission Control Protocol/Internet Protocol. A networking system that uses specific station addresses to route packets. TCP/IP allows hams to send messages or exchange files over great distances with a high degree of reliability.

terminal: A device that allows a human operator to communicate with a computer or *TNC*. Also known as a *data terminal*.

terminal node controller: (*TNC*) a device that assembles packets for transmission and decodes received packets for display on computers or *data terminals*.

terminal software: software that allows a personal computer to imitate (emulate) the functions of a *data terminal*.

TexNet: A centralized, high-speed packet network that originated in Texas. It can now be found in other states as well.

TheNet: Another version of *NET/ROM* node-networking software.

TNC: an abbreviation for *terminal node controller*.

TTL: a type of digital interface based on transistor-transistor logic. For example, some personal computers offer a TTL interface rather than an EIA-232-E interface.

unconnected packet: a packet transmitted without a specific destination. Used for beacons, calling CQ and roundtable communications.

uplink: in amateur satellite communications, a frequency used to send information to a satellite.

upload: to send a file or message to a *PBBS*, satellite or other Amateur Radio packet station.

weather node: a *node* designed to provide weather data to several users at once.

X.25: a packet networking protocol designed for use over telephone lines. The Amateur *AX.25* protocol is based on X.25.

Packet Info Guide

Packet Info Guide

Unless otherwise indicated, all items in this section may be purchased, subject to availability, from your dealer or from the American Radio Relay League, 225 Main St, Newington, CT 06111; tel 203-666-1541. Prices subject to change without notice.

Books

Your Gateway to Packet Radio by Stan Horzepa, WA1LOU. A thorough discussion of packet radio, including many of the latest developments. ARRL Order No. 2030, $12

The ARRL Operating Manual. Stan Horzepa, WA1LOU, addresses operating techniques and various networking systems such as NET/ROM, TCP/IP, ROSE and TexNet. ARRL Order No. 1086, $18

AX.25 Amateur Packet-Radio Link-Layer Protocol—Learn the inner workings of packet protocol. ARRL Order No. 0119, $8

Satellite Experimenter's Handbook—By Martin Davidoff, K2UBC. The ultimate reference for the satellite operator. ARRL Order No. 3185, $20.

Decoding Telemetry from Amateur Satellites—G. Gould Smith, WA4SXM, shows you how to receive and decode telemetry signals. Available from AMSAT, PO Box 27, Washington, DC 20044; tel 301-589-6062. $15

Newsletters

Packet Status Register—Published quarterly by Tucson Amateur Packet Radio (TAPR), 8987-309 E Tanque Verde Rd, #337, Tucson, AZ 85749-9399. $15/year.

The AMSAT Journal—available from AMSAT, PO Box 27, Washington, DC 20044; tel 301-589-6062. $30/year.

OSCAR Satellite Report—available from R. Myers Communications, PO Box 17108, Fountain Hill, AZ 85269-7108. $56/year US, $62/year Canada.

PBBS Software

A variety of PBBS and mailbox software is available for most popular personal computers. Here is a list of the software available as of the publication date of this book.

Atari 520ST/1040ST

Atari ST Mailbox—A WØRLI-type mailbox program for the Atari 520ST and 1040ST computers. To obtain a copy of the software, contact Thor Andersen, LA2DAA, at Riddersporen 6, N-3032 Drammen, Norway.

PBBS—Mike Curtis, WD6EHR, has ported the WØRLI PBBS software to the Atari 520ST and 1040ST computers. The program has most of the features of the original program and is available by sending a blank $3^1/_2$-inch diskette and postpaid diskette mailer. Contact Mike at 7921 Wilkinson Ave, N Hollywood, CA 91605.

Commodore Amiga

AmigaBBS—Randal Lilly, N3ET, wrote this PBBS software to run with the Kantronics KPC-1, KPC-2, KPC-4 and KAM TNCs. The software is available on the Amiga Amateur Radio Public Domain Disk #5, which is distributed by Kathy Wehr, WB3KRN, RD#1, Box 193, Watsontown, PA 17777. To obtain a copy of the disk, send a blank formatted disk, a sturdy return envelope, a label printed with your name and address and sufficient return postage.

Commodore C64

C64 Packet Talker—This unique mailbox system for the C64 stores messages for up to 300 users and converts all packet-radio messages to voice. It is available from Engineering Consulting, 583 Candlewood St, Brea, CA 92621.

Packet Radio BBS—This C64 program was written in BASIC by Verne Buland, W9ZGS. It may be downloaded from CompuServe's HamNet.

WB4APR PBBS—Bob Bruninga, WB4APR, wrote Commodore C-64 software that emulates many of the features of the WØRLI PBBS. The features include message storage, limited file storage and automatic message forwarding. This BASIC program may be obtained by sending a formatted diskette and $5 to Bob at 59 Southgate Ave, Annapolis, MD 21401.

IBM PCs and Compatibles

ARES/Data—ARES/Data is a multiple-connection, multiport database and conference bridge designed specifically for tracking victims and emergency personnel in a disaster. It was written by William Moerner, WN6I, and David Palmer, N6KL, and may be downloaded from CompuServe's HamNet.

BB—The BB program for the IBM PC and compatibles is a multiple-connection PBBS that was written by Roy Engehausen, AA4RE. It may be downloaded from CompuServe's HamNet, or from the WA6RDH BBS at 916-678-1535. It is also available on disk from the Tucson Amateur Packet Radio Corporation (TAPR), 8987-309 E Tanque Verde Rd, #337 Tucson, AZ 85749-9399.

CBBS Mailbox System—CBBS was written for the IBM PC and compatibles by Ed Picchetti, K3RLI, and Joe Lagermasini, AG3F. It may be downloaded from CompuServe's HamNet and is also available on disk from the Tucson Amateur Packet Radio Corporation (TAPR), 8987-309 E Tanque Verde Rd, #337 Tucson, AZ 85749-9399.

MPC—This multiport AMTOR and packet-radio BBS for the IBM PC and compatibles was written by Lacy McCall, AC4X. It may be downloaded from CompuServe's HamNet.

MSYS—MSYS is a multiple-user, multiport PBBS that runs on the IBM PC and compatibles (it requires a hard disk). MSYS supports gateway, KaNode and TCP/IP operations and was written by Mike Pechura, WA8BXN. It may be downloaded from CompuServe's HamNet.

ROSErver/Packet Radio Mailbox System (PRMBS)—This multiple-user PBBS/packet server for the IBM PC and compatibles attempts to eliminate some of the SYSOP maintenance that other systems require. Written by Brian Riley, KA2BQE, and Dave Trulli, NN2Z, it may be downloaded from the RATS BBS at 201-387-8898 and is available on disk from the Tucson Amateur Packet Radio Corporation (TAPR), 8987-309 E Tanque Verde Rd, #337 Tucson, AZ 85749-9399.

WØRLI Mailbox—This is the original PBBS software rewritten in C by Hank Oredson, WØRLI, and David Toth, VE3GYQ. The current version is intended for the IBM PC and compatibles and may be downloaded from CompuServe's HamNet, the WA6RDH BBS (916-678-1535), the VE3GYQ BBS (active at 519-660-1442 when a new version is out) and the VE4UB BBS (204-785-8518), and is also available on disk from the Tucson Amateur Packet Corporation (TAPR), 8987-309 E Tanque Verde Rd, #337 Tucson, AZ 85749-9399.

WA7MBL PBBS—This is the Jeff Jacobsen, WA7MBL, implementation of the original WØRLI PBBS for the IBM PC and compatibles. It is fully compatible with WØRLI systems. The WA7MBL PBBS software may be downloaded from CompuServe's HamNet and is available on disk from the Tucson Amateur Packet Radio Corporation (TAPR), 8987-309 E Tanque Verde Rd, #337 Tucson, AZ 85749-9399.

Wake Digital Communications Group (WDCG) PBBS—The WDCG's PBBS runs on the IBM PC and compatibles. It supports file transfers using ASCII, XMODEM and XPACKET protocols. The system includes a fully developed message bulletin board, but does not support mail forwarding. The software may be downloaded from CompuServe's HamNet or may be obtained from the WDCG, c/o Randy Ray, WA5SZL, 9401 Taurus Ct, Raleigh, NC 27612.

Tandy Color Computer

WJ5W CoCo/PBBS—Monty W. Haley, WJ5W, produced a partial implementation of the WØRLI PBBS for the Radio Shack/Tandy Color Computer. The program is written in BASIC and may be obtained directly from Monty at Rte 1, Box 210-B, Evening Shade, AR 72532.

Packet Equipment Manufacturers

A & A Engineering, 2521 West La Palma, Suite K, Anaheim, CA 92801, tel 714-952-2114.

Advanced Electronic Applications (AEA), PO Box C2160, Lynnwood, WA 98036-0918, tel 206-774-5554.

Gracilis, 623 Palace St, Aurora, IL 60506, tel 708-897-9346.

HAL Communications Corp, PO Box 365, Urbana, IL 61801, tel 217-367-7373.

Kantronics, 1202 E 23rd St, Lawrence, KS 66046, tel 913-842-7745.

L. L. Grace Corporation, 41 Acadia Dr, Voorhees, NJ 08043, tel 609-751-1018.

MFJ Enterprises, Box 494, Mississippi State, MS 39762, tel 601-323-5869.

PacComm, 4413 N Hesperides St, Tampa, FL 33614-7618, tel 813-874-2980.

Satellite Software

Software for satellite tracking, telemetry decoding and Pacsat operation is available from: AMSAT, PO Box 27, Washington, DC 20044; tel 301-589-6062. Send a self-addressed, stamped envelope and ask for their software catalog.

AMSAT-NA Digital Satellite Guide—Includes the *PB/PG* software required to access many of the Pacsats. Available from AMSAT at the above address. $12 US, $13 Canada/Mexico, $15 elsewhere.

WISP—A *Windows*-based program which performs the same functions as the *PB/PG* software mentioned above. Available from AMSAT at the above address.

DISPLAY4—A freeware program designed specifically for viewing UoSAT-OSCAR 22 and KITSAT-OSCAR 23 images. A VGA or Super VGA monitor is required. The software is available on many Amateur-Radio oriented telephone bulletin boards and on CompuServe's HAMNET. Check the following telephone BBSs for availability:

Dallas Remote Imaging Group (DRIG)—tel 214-394-7438

ARRL—tel 203-666-0578

N8EMR—tel 614-895-2553

Satellite Modems

Tucson Amateur Packet Radio, 8987-309 E Tanque Verde Rd, #337 Tucson, AZ 85749-9399; tel 817-383-0000.

PacCOMM Inc, 4413 N Hesperides St, Tampa, FL 33614-7618; tel 813-874-2980.

Satellite Orbital Elements

Orbital elements for all active Amateur Radio satellites are published in amateur satellite newsletters (see above). They are also transmitted twice weekly by W1AW. See the W1AW schedule in this section.

In addition, orbital elements are available on AMSAT Information Nets

International: Sunday at 1800 UTC on 14.282 MHz and 2100 UTC on 21.280 MHz.

International: Saturday at 2200 UTC on 18.155 MHz

US East Coast: Tuesday at 2100 EST/EDT on 3.840 MHz

US Central: Tuesday at 2100 CST/CDT on 3.840 MHz

US West Coast: Tuesday at 2000 PST/PDT on 3.840 MHz

TCP/IP Address Coordinators

Note: Coordinators are listed in order of their geographic areas. This list was up to date as of September 1, 1994. It is subject to change without notice.

State	Coordinator
AK	John Stannard, KL7JL
AL	Richard Elling, KB4HB
AR	Richard Duncan, WD5B
AZ	David Dodell, WB7TPY
CA (Antelope Valley/ Kern County)	Dana Myers, KK6JQ
CA (Los Angeles— San Fernando Valley)	Jeff Angus, WA6FWI
CA (Orange County)	Terry Neal, AA6TN
CA (Sacramento)	Bob Meyer, K6RTV
CA (Santa Barbara/Ventura)	Don Jacob, WB5EKU

CA (San Bernardino and Riverside)	Geoffrey Joy, KE6QH
CA (San Diego)	Brian Kantor, WB6CYT
CA (Silicon Valley— San Francisco)	Douglas Thom, N6OYU
CO (northeast)	Fred Schneider, KØYUM
CO (southeast)	Bdale Garbee, N3EUA
CO (western)	Bob Ludtke, K9MWM
CT	Bill Lyman, N1NWP
DC	Don Bennett, K4NGC
DE	John DeGood, NU3E
FL	Bruce La Pointe, WD4HIM
GA	Doug Drye, KD4NC
HI and Pacific islands	John Shalamskas, KJ9U
ID	Steven King, KD7RO
IL (central and southern)	Chuck Henderson, WB9UUS
IL (northern)	Ken Stritzel, WA9AEK
KS	Dale Puckett, KØHYD
KY	Tyler Barnett, N4TY
LA	James Dugal, N5KNX
MA (center and eastern)	Johnathan Vail, N1DXG
MA (western)	Bob Wilson, KA1XN
MD	Howard Leadmon, WB3FFV
ME	Carl Ingerson, N1DXM
MI (upper peninsula)	Pat Davis, KD9UU
MI (lower peninsula)	Jeff King, WB8WKA
MN	Gary Sharp, WDØHEB
MN (Minneapolis only)	Andy Warner, NØREN
MO	Stan Wilson, AKØB
MS	John Martin, KB5GGO
MT	Steven Elwood, N7GXP
NC (eastern)	Mark Bitterlich, WA3JPY

NC (western)	Charles Layno, WB4WOR
ND	Steven Elwood, N7GXP
NE	Mike Nickolaus, NFØN
NH	Gary Grebus, K8LT
NJ (northern)	Dave Trulli, NN2Z
NJ (southern)	Bob Applegate, WA2ZZX
NM	J Gary Bender, WS5N
NY (eastern)	Bob Bellini, N2IGU
NY (western)	Paul Gerwitz, WA2WPI
New York City and Long Island	Bob Foxworth, K2EUH
NV (southern)	Earl Petersen, KF7TI
NV (northern)	Bill Healy, N8KHN
OH	Gary Sanders, N8EMR
OK	Joe Buswell, K5JB
OR	Ron Henderson, WA7TAS
OR (northwest and Vancouver, WA)	Tom Kloos, WS7S
PA (eastern)	Doug Crompton, WA3DSP
PA (western)	Bob Hoffman, N3CVL
PR	Karl Wagner, KP4QG
RI	Charles Greene, W1CG
SC	Mike Abbott, N4QXV
SD	Steven Elwood, N7GXP
TN	Mark J. Bailey, N4XHX
TX (northern)	Don Adkins, KD5QN
TX (southern)	Kurt Freiberger, WB5BBW
TX (western)	Rod Huckabay, KA5EJX
UT	Matt Simmons, KG7MH
VA	Jim DeArras, WA4ONG
VA (Charlottesville)	Jon Gefaell, KD4CQY

VI	Bernie McDonnell, NP2W
VT	Ralph Stetson, KD1R
WA (eastern)	Steven King, KD7RO
WA (western)	Dennis Goodwin, KB7DZ
WI	Pat Davis, KD9UU
WV	Rich Clemens, KB8AOB
WY	Reid Fletcher, WB7CJO

TCP/IP Software

There is typically a nominal fee charged to cover the cost of disks and postage, so send an SASE first to find out what is required. Also, some versions of TCP/IP software can be downloaded from CompuServe's HamNet Library 9.

Apple Macintosh: Doug Thom, N6OYU, c/o Thetherless Access Ltd, 1405 Graywood Dr, San Jose, CA 95129-2210.

Atari ST: Mike Curtis, WD6EHR, 7921 Wilkinson Ave, North Hollywood, CA 91605-2210.

Commodore Amiga: John Heaton, G1YYH, MCC Network Unit, Oxford Rd, Manchester M13 9PL, United Kingdom.

IBM-PCs and compatibles: Tucson Amateur Packet Radio, 8987-309 E Tanque Verde Rd, #337 Tucson, AZ 85749-9399.

W1AW Schedule

Pacific	Mtn	Cent	East	Sun	Mon	Tue	Wed	Thu	Fri	Sat
6 am	7 am	8 am	9 am			Fast Code	Slow Code	Fast Code	Slow Code	
7 am	8 am	9 am	10 am			Code Bulletin				
8 am	9 am	10 am	11 am			Teleprinter Bulletin				
9 am	10 am	11 am	noon							
10 am	11 am	noon	1 pm			**Visiting Operator Time**				
11 am	noon	1 pm	2 pm							
noon	1 pm	2 pm	3 pm							
1 pm	2 pm	3 pm	4 pm	Slow Code	Fast Code	Slow Code	Fast Code	Slow Code	Fast Code	Slow Code
2 pm	3 pm	4 pm	5 pm	Code Bulletin						
3 pm	4 pm	5 pm	6 pm	Teleprinter Bulletin						
4 pm	5 pm	6 pm	7 pm	Fast Code	Slow Code	Fast Code	Slow Code	Fast Code	Slow Code	Fast Code
5 pm	6 pm	7 pm	8 pm	Code Bulletin						
6 pm	7 pm	8 pm	9 pm	Teleprinter Bulletin						
6^{45} pm	7^{45} pm	8^{45} pm	9^{45} pm	Voice Bulletin						
7 pm	8 pm	9 pm	10 pm	Slow Code	Fast Code	Slow Code	Fast Code	Slow Code	Fast Code	Slow Code
8 pm	9 pm	10 pm	11 pm	Code Bulletin						
9 pm	10 pm	11 pm	Mdnte	Teleprinter Bulletin						
9^{45} pm	10^{45} pm	11^{45} pm	12^{45} am	Voice Bulletin						

Note: W1AW's schedule is at the same local time throughout the year. The schedule according to your local time will change if your local time does not have seasonal adjustments that are made at the same time as North American time changes between standard time and daylight time. From the first Sunday in April to the last Sunday in October, UTC = Eastern Time + 4 hours. For the rest of the year, UTC = Eastern Time + 5 hours.

❏ *Morse code transmissions:*
Frequencies are 1.818, 3.5815, 7.0475, 14.0475, 18.0975, 21.0675, 28.0675 and 147.555 MHz.
Slow Code = practice sent at 5, 7^{1}/$_{2}$, 10, 13 and 15 wpm.
Fast Code = practice sent at 35, 30, 25, 20, 15, 13 and 10 wpm.
Code practice text is from the pages of *QST*. The source is given at the beginning of each practice session and alternate speeds within each session. For example, "Text is from July 1992 *QST*, pages 9 and 81," indicates that the plain text is from the article on page 9 and mixed number/letter groups are from page 81.
Code bulletins are sent at 18 wpm.

❏ *Teleprinter transmissions:*
Frequencies are 3.625, 7.095, 14.095, 18.1025, 21.095, 28.095 and 147.555 MHz.
Bulletins are sent at 45.45-baud Baudot and 100-baud AMTOR, FEC Mode B.
110-baud ASCII will be sent only as time allows.
On Tuesdays and Saturdays at 6:30 PM Eastern Time, Keplerian elements for many amateur satellites are sent on the regular teleprinter frequencies.

❏ *Voice transmissions:*
Frequencies are 1.855, 3.99, 7.29, 14.29, 18.16, 21.39, 28.59 and 147.555 MHz.

❏ *Miscellanea:*
On Fridays, UTC, a DX bulletin replaces the regular bulletins.
W1AW is open to visitors during normal operating hours: from 1 PM until 1 AM on Mondays, 9 AM until 1 AM Tuesday through Friday, from 1 PM to 1 AM on Saturdays, and from 3:30 PM to 1 AM on Sundays. FCC licensed amateurs may operate the station from 1 to 4 PM Monday through Saturday. Be sure to bring your current FCC amateur license or a photocopy.
In a communications emergency, monitor W1AW for special bulletins as follows: voice on the hour, teleprinter at 15 minutes past the hour, and CW on the half hour.
Headquarters and W1AW are closed on New Year's Day, President's Day, Good Friday, Memorial Day, Independence Day, Labor Day, Thanksgiving and the following Friday, and Christmas Day. On the first Thursday of September, Headquarters and W1AW will be closed during the afternoon.

11

About The American Radio Relay League

The seed for Amateur Radio was planted in the 1890s, when Guglielmo Marconi began his experiments in wireless telegraphy. Soon he was joined by dozens, then hundreds, of others who were enthusiastic about sending and receiving messages through the air—some with a commercial interest, but others solely out of a love for this new communications medium. The United States government began licensing Amateur Radio operators in 1912.

By 1914, there were thousands of Amateur Radio operators—hams—in the United States. Hiram Percy Maxim, a leading Hartford, Connecticut, inventor and industrialist saw the need for an organization to band together this fledgling group of radio experimenters. In May 1914 he founded the American Radio Relay League (ARRL) to meet that need.

Today ARRL, with more than 160,000 members, is the largest organization of radio amateurs in the United States. The League is a not-for-profit organization that:

•promotes interest in Amateur Radio communications and experimentation

•represents US radio amateurs in legislative matters, and

•maintains fraternalism and a high standard of conduct among Amateur Radio operators.

At League headquarters in the Hartford suburb of Newington, the staff helps serve the needs of members. ARRL is also International Secretariat for the International Amateur Radio Union, which is made up of similar societies in more than 100 countries around the world.

ARRL publishes the monthly journal *QST*, as well as newsletters and many publications covering all aspects of Amateur Radio. Its headquarters station, W1AW, transmits bulletins of interest to radio amateurs and Morse code practice sessions. The League also coordinates an extensive field organization, which includes volunteers who provide technical information for radio amateurs and public-service activities. ARRL also represents US amateurs with the Federal Communications Commission and other government agencies in the US and abroad.

Membership in ARRL means much more than receiving *QST* each month. In addition to the services already described, ARRL offers membership services on a personal level, such as the ARRL Volunteer Examiner Coordinator Program and a QSL bureau.

Full ARRL membership (available only to licensed radio amateurs) gives you a voice in how the affairs of the organization are governed. League policy is set by a Board of Directors (one from each of 15 Divisions). Each year, half of the ARRL Board of Directors stands for election by the full members they represent. The day-to-day operation of ARRL HQ is managed by an Executive Vice President and a Chief Financial Officer.

No matter what aspect of Amateur Radio attracts you, ARRL membership is relevant and important. There would be no Amateur Radio as we know it today were it not for the ARRL. We would be happy to welcome you as a member! (An Amateur Radio license is not required for Associate Membership.) For more information about ARRL and answers to any questions you may have about Amateur Radio, write or call:

ARRL Educational Activities Dept
225 Main Street
Newington CT 06111-1494
(203) 666-1541

Prospective new amateurs call:
800-32-NEW HAM (800-326-3942)

Index

(Note: "INFO" refers to the Packet Info Guide.)

169

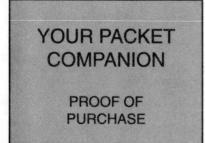

YOUR PACKET
COMPANION

PROOF OF
PURCHASE

Please use this form to give us your comments on this book and what you'd like to see in future editions.

Where did you purchase this book?
  □ From ARRL directly   □ From àn ARRL dealer

Is there a dealer who carries ARRL publications within:
  □ 5 miles   □ 15 miles   □ 30 miles   of your location?   □ Not sure.

License class:
  □ Novice   □ Technician   □ Technician with HF privileges
  □ General   □ Advanced   □ Extra

Name	ARRL member? □ Yes □ No
_____	Call sign _____
Daytime Phone ( ) _____	Age _____

Address _____

City, State/Province, ZIP/Postal Code_____

If licensed, how long? _____

Other hobbies_____

_____

Occupation _____

For ARRL use only   Pkt Comp
Edition 1 2 3 4 5 6 7 8 9 10 11 12
Printing   3 4 5 6 7 8 9 10 11 12

From _____

_____

_____

EDITOR, YOUR PACKET COMPANION
AMERICAN RADIO RELAY LEAGUE
225 MAIN ST
NEWINGTON CT 06111

····················· please fold and tape ·····················